Springer Theses

Recognizing Outstanding Ph.D. Research

Aims and Scope

The series "Springer Theses" brings together a selection of the very best Ph.D. theses from around the world and across the physical sciences. Nominated and endorsed by two recognized specialists, each published volume has been selected for its scientific excellence and the high impact of its contents for the pertinent field of research. For greater accessibility to non-specialists, the published versions include an extended introduction, as well as a foreword by the student's supervisor explaining the special relevance of the work for the field. As a whole, the series will provide a valuable resource both for newcomers to the research fields described, and for other scientists seeking detailed background information on special questions. Finally, it provides an accredited documentation of the valuable contributions made by today's younger generation of scientists.

Theses are accepted into the series by invited nomination only and must fulfill all of the following criteria

- They must be written in good English.
- The topic should fall within the confines of Chemistry, Physics, Earth Sciences, Engineering and related interdisciplinary fields such as Materials, Nanoscience, Chemical Engineering, Complex Systems and Biophysics.
- The work reported in the thesis must represent a significant scientific advance.
- If the thesis includes previously published material, permission to reproduce this must be gained from the respective copyright holder.
- They must have been examined and passed during the 12 months prior to nomination.
- Each thesis should include a foreword by the supervisor outlining the significance of its content.
- The theses should have a clearly defined structure including an introduction accessible to scientists not expert in that particular field.

More information about this series at http://www.springer.com/series/8790

Kenji Yasuda

Emergent Transport Properties of Magnetic Topological Insulator Heterostructures

Doctoral Thesis accepted by The
University of Tokyo, Tokyo, Japan

 Springer

Author
Dr. Kenji Yasuda
Department of Physics
Massachusetts Institute of Technology
Cambridge, MA, USA

Supervisor
Prof. Yoshinori Tokura
RIKEN Center for Emergent
Matter Science (CEMS)
Wako, Japan

The University of Tokyo
Tokyo, Japan

ISSN 2190-5053   ISSN 2190-5061   (electronic)
Springer Theses
ISBN 978-981-15-7185-5   ISBN 978-981-15-7183-1   (eBook)
https://doi.org/10.1007/978-981-15-7183-1

This Springer imprint is published by the registered company Springer Nature Singapore Pte Ltd.
The registered company address is: 152 Beach Road, #21-01/04 Gateway East, Singapore 189721,
Singapore

Supervisor's Foreword

The application of quantum mechanics to condense matter has made a significant impact in understanding and controlling solid-state materials over the last century. Despite the huge success of semiconductor devices in modern electronics, they do not harness the full potential of quantum mechanical wave functions. Namely, they utilize only the energy-momentum dispersion of the band structure but not explicitly the phase information of the wave function. Over the past few decades, people have started to appreciate the importance of the geometrical phases and topology in momentum space motivated primarily by quantum Hall effect and anomalous Hall effect. The concept of topology now forms one of the central schemes in condensed matter physics with the discovery of novel quantum materials, such as topological insulators. The exotic band topology of these quantum materials is expected to serve as a novel platform for emergent physical properties.

This thesis by Dr. Kenji Yasuda presents such possibilities by studying the electrical transport properties of topological insulator heterostructures. Employing state-of-the-art thin film growth and fabrication techniques, he has found emergent transport phenomena; dissipationless chiral edge conduction along the magnetic domain wall, skyrmion formation, current-induced magnetization switching, and nonreciprocal nonlinear resistance, i.e., diode-like effect, in magnetic and super-conducting topological insulators. These responses all result from highly entangled electron spin and momentum enabled by the topological nature of the band structure. The unique spin-momentum locking of the topological surface state, coupled with magnetism, enabled the electrical control of magnetism and vice versa. The present study provides superior insight into the transport phenomena of topological materials, offering a guideline for utilizing topological band structure to produce novel physical functionalities.

Wako/Tokyo, Japan   Prof. Yoshinori Tokura
August 2020

Preface

One of the central themes of condensed matter physics is to understand and utilize the characteristics of phases of matter. Roughly speaking, there are two types of phases in condensed matter: one is a spontaneous symmetry broken ordered phase, and the other is a topological phase of matter. The notion of spontaneous symmetry breaking leads to the appearance of a variety of orders such as ferroelectricity, ferromagnetism and superconductivity with fixed order parameters. These effects were experimentally discovered and theoretically formulated by the 1950s and have long been the central theme of condensed matter physics ever since. The concept of topological phase, on the other hand, is relatively new. The most prominent example of the topological phase is the quantum Hall effect discovered in the 1980s in two-dimensional electronic systems under the magnetic field. Thanks to this discovery, the importance of band topology in the Brillouin zone is recognized for the first time. From this moment, the idea of topology originally developed in pure mathematics began to prevail in condensed matter physics. Followed by the formulation of anomalous Hall effect and spin Hall effect, the extension of quantum Hall effect has lead to the discovery of various topological materials such as topological insulator, Dirac semimetal and Weyl semimetal. These materials are characterized by a topological integer such as winding number, Chern number, and Z_2 invariant. Since these values are integers, the continuous deformation cannot change the topological value. As a result, these materials yield characteristic surface state at the interface between the vacuum or trivial materials, associated with the discrete change of the topological integer. Such surface states gained interests of physicists because of their robustness against perturbations as well as the characteristic band structures that cannot be realized in conventional bulk materials. The surface state possesses a spin-nondegenerate massless Dirac dispersion, where its spin direction is uniquely locked to its momentum called spin-momentum locking. Following the theoretical prediction, the experimental study of topological insulators has developed rapidly over the past 10 years. Together with material developments, the detailed characteristics of the surface state are now well understood.

Now that the respective characteristics of the two phases, symmetry broken ordered phase and topological phase are well understood, it is natural to expect that the combination of these two will result in more exotic quantum states. A naive consideration can readily predict the appearance of exotic phenomena: When magnetization is combined with the spin-momentum locked topological surface state, the magnetization dramatically modifies the band structure giving rise to various magnetoelectric responses. In fact, quantum anomalous Hall effect, or the magnetically induced quantum Hall effect, and various spintronic phenomena are experimentally found in a magnetically coupled topological insulator. Meanwhile, when superconductivity is combined with spin-momentum locking, the exotic pairing is expected because of the limited pairing symmetry by the spin momentum locking. This results in the emergence of topological superconductivity and associated formation of Majorana fermion.

Although there have been a lot of seminal works in this direction from around 2010, this field is still immature. The goal of the thesis is to investigate the emergent transport properties and functionalities of topological insulators under broken symmetry. To study the interaction between magnetization and surface state, we study the transport properties of $Cr_x(Bi_{1-y}Sb_y)_{2-x}Te_3$ and $(Bi_{1-y}Sb_y)_2Te_3/Cr_x$ $(Bi_{1-y}Sb_y)_{2-x}Te_3$ the research has found the high-temperature realization of the quantum Hall effect and the chiral edge state at the domain wall in quantum anomalous Hall state. In addition, we found various spintronic functionalities such as unidirectional magnetoresistance, current-nonlinear Hall effect, current-induced magnetization switching, and skyrmion formation. Finally, the interaction between surface state and superconductivity is investigated in the heterostructure of Bi_2Te_3 and FeTe. The largely enhanced nonreciprocal transport is found associated with the appearance of superconductivity.

Cambridge, USA   Dr. Kenji Yasuda

List of publications

Parts of this thesis have been published in the following journal articles:

1. **Quantum Hall states stabilized in semi-magnetic bilayers of topological insulators**
   R. Yoshimi*, K. Yasuda*, A. Tsukazaki, K. S. Takahashi, N. Nagaosa, M. Kawasaki and Y. Tokura
   Nature Communications **6**, 8530 (2015) (* equal contribution).
2. **Magnetic modulation doping in topological insulators toward higher-temperature quantum anomalous Hall effect**
   M. Mogi, R. Yoshimi, A. Tsukazaki, K. Yasuda, Y. Kozuka, K. S. Takahashi, M. Kawasaki and Y. Tokura
   Applied Physics Letters **107**, 182401 (2015).
3. **Geometric Hall effects of topological insulator heterostructures**
   K. Yasuda, R. Wakatsuki, T. Morimoto, R. Yoshimi, A. Tsukazaki, K. S. Takahashi, M. Ezawa, M. Kawasaki, N. Nagaosa and Y. Tokura
   Nature Physics **12**, 555–559 (2016).
4. **Zero-bias photocurrent in ferromagnetic topological insulator**
   N. Ogawa, R. Yoshimi, K. Yasuda, A. Tsukazaki, M. Kawasaki and Y. Tokura
   Nature Communications **7**, 12246 (2016).
5. **Large Unidirectional Magnetoresistance in a Magnetic Topological Insulator**
   K. Yasuda, A. Tsukazaki, R. Yoshimi, K. S. Takahashi, M. Kawasaki and Y. Tokura
   Physical Review Letters **117**, 127202 (2016).
6. **Current-Nonlinear Hall Effect and Spin-Orbit Torque Magnetization Switching in a Magnetic Topological Insulator**
   K. Yasuda, A. Tsukazaki, R. Yoshimi, K. Kondou, K. S. Takahashi, Y. Otani, M. Kawasaki and Y. Tokura
   Physical Review Letters **119**, 137204 (2017).
7. **Quantized chiral edge conduction on domain walls of a magnetic topological insulator**
   K. Yasuda, M. Mogi, R. Yoshimi, A. Tsukazaki, K. S. Takahashi, M. Kawasaki, F. Kagawa and Y. Tokura
   Science **358**, 1311–1314 (2017).
8. **Nonreciprocal charge transport at topological insulator/superconductor interface**
   K. Yasuda, H. Yasuda, R. Yoshimi, A. Tsukazaki, K. S. Takahashi, N. Nagaosa, M. Kawasaki, Y. Tokura
   Nature Communications **10**, 2734 (2019).
9. **Magnetic topological insulators**
   Y Tokura, K. Yasuda, A Tsukazaki
   Nature Reviews Physics **1**, 126–143 (2019).

Some related works, which not presented here, are published in the following articles:

- **Interface-driven topological Hall effect in SrRuO$_3$-SrIrO$_3$ bilayer**
  J. Matsuno, N. Ogawa, K. Yasuda, F. Kagawa, W. Koshibae, N. Nagaosa, Y. Tokura, M. Kawasaki
  Science Advances **2**, e1600304 (2016).
- **Current-driven magnetization switching in ferromagnetic bulk Rashba semiconductor (Ge,Mn)Te**
  R. Yoshimi, K. Yasuda, A. Tsukazaki, K. S. Takahashi, M. Kawasaki, Y. Tokura
  Science Advances **4**, eaat9989 (2018).

Acknowledgements

I would like to express my sincerest gratitude to Prof. Yoshinori Tokura for his continuous encouragement and helpful research guidance throughout this work.

I would like to appreciate especially the following people for their valuable collaboration: Prof. M. Kawasaki, Prof. N. Nagaosa, Prof. A. Tsukazaki, Prof. F. Kagawa, Dr. K. S. Takahashi, Dr. R. Yoshimi, Dr. Y. Kozuka, Dr. T. Morimoto, Dr. R. Wakatsuki, Prof. M. Ezawa, Prof. Y. Otani, Dr. K. Kondou, Dr. N. Ogawa, Dr. M. Kawamura, Dr. M. Mogi, and Mr. H. Yasuda.

I am grateful to Prof. Naoto Nagaosa, Prof. Yoshihiro Iwasa, Prof. Eiji Saito, and Prof. Takasada Shibauchi for their fruitful comments and recommendations on this thesis. I also thank all members of the Tokura group in the University of Tokyo and of RIKEN center for emergent matter science (CEMS) for their kind encouragement and experimental help through my research life.

Finally, I would like to express my special appreciation to my family for their warm encouragement and support for years. I dedicate this work to my wife, who has been a constant source of encouragement throughout the graduate studies.

April 2020

Dr. Kenji Yasuda

Contents

Chapter 1
Introduction

1.1 Theoretical Background of Topological Insulator

1.1.1 Quantum Hall Effect (QHE)

The importance of the band topology in momentum space was recognized by the discovery of the Quantum Hall effect (QHE). Here, we introduce the fundamental physics of QHE before the explanation of topological insulators.

In the 1980s, QHE is observed in two-dimensional electron gas (2DEG) under magnetic field such as in Si-MOSFET and at GaAs/AlGaAs interface. Here, the Hall conductivity takes a quantized value, and simultaneously, longitudinal conductivity becomes zero [1]. Figure 1.1 is the first observation of QHE by von. Klitzing. The Hall voltage takes a plateau at a specific gate voltage, and the longitudinal voltage goes to zero. Especially, the Hall conductivity is quantized to

$$\sigma_{xy} = Ne^2/h, \tag{1.1}$$

where N takes the integer value.

After the experimental discovery of QHE, it is theoretically formulated by Thouless, Khomoto, Nightnagle, den Nijs (TKNN), which is derived from the Kubo formula [2]. According to the TKNN formula, the two-dimensional electron system with broken time reversal symmetry is characterized by the topological invariant called TKNN number (Chern number) ν (integer). Here, the system Hall conductivity σ_{xy} can be calculated as follows:

$$\sigma_{xy} = \nu e^2/h, \tag{1.2}$$

$$\nu = \sum_{n:\text{filled}} \int_{\text{BZ}} \frac{d^2k}{2\pi} \left(\frac{\partial a_{n,y}}{\partial k_x} - \frac{\partial a_{n,x}}{\partial k_y} \right), \tag{1.3}$$

$$a_n(\mathbf{k}) = -i \langle u_{nk} \mid \partial_k \mid u_{nk} \rangle. \tag{1.4}$$

© Springer Nature Singapore Pte Ltd. 2020
K. Yasuda, *Emergent Transport Properties of Magnetic Topological
Insulator Heterostructures*, Springer Theses,
https://doi.org/10.1007/978-981-15-7183-1_1

Fig. 1.1 The gate voltage dependence of transport property in Si-MOSFET under $B = 18$ T. U_{pp} is the longitudinal voltage and U_H is the Hall voltage. Reprinted figure with permission from [1] Copyright 1980 by the American Physical Society

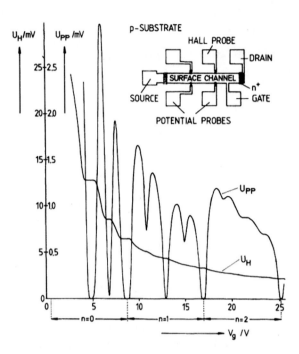

Here, $a_n(k)$ is a Berry connection and corresponds to the twisting of the wave function. Since ν is an integer number, it cannot change with the continuous deformation of the material parameters so that it can never be continuously connected to the trivial insulator $\nu = 0$ or a vacuum. As a result, the gapless edge state inevitably appears at the sample edge. This concept is called a bulk-boundary correspondence.

1.1.2 Z_2 Invariant and Topological Insulator

The topological invariant in TKNN formula, defined in the time reversal symmetry broken two-dimensional electron systems, is characterized by an integer group Z. On the other hand, the time reversal invariant 2D or 3D system is shown to be characterized by a Z_2 topological invariant during 2005–2007 [3–7]. Here, Z_2 is the cycling group of order two and has $\{0, 1\}$ as its element. In other words, the quantum Hall effect is characterized by the number of edge states, while topological insulator is characterized by the existence or the absence of the surface states. In the following, we introduce the Z_2 invariant in the two-dimensional system and expand it to the three-dimensional system.

Z_2 Invariant in a Two-dimensional System

First, we construct the matrix w for all the occupied states as follows:

$$w_{\alpha\beta}(\boldsymbol{k}) = \langle u_{\alpha,-\boldsymbol{k}}|\Theta|u_{\beta,\boldsymbol{k}}\rangle, \tag{1.5}$$

$$\Theta = -is_y K, \tag{1.6}$$

where Θ is a time reversal operator. As a result, the matrix element becomes finite when the pair of bands are connected by time reversal operation (Kramers pair). The calculation of the following quantity by the matrix w gives us the classification of Z_2 topology.

$$(-1)^{\nu} = \prod_{i=1}^{4} \frac{\mathrm{Pf}[w(\Lambda_i)]}{\sqrt{\det[w(\Lambda_i)]}}. \tag{1.7}$$

Here, Λ_i is called a time reversal invariant momentum (TRIM) and is a momentum where $\boldsymbol{k}$ and $-\boldsymbol{k}$ becomes identical in the Brillouin zone. For example, the four points $(0,0)$, $(\pi,0)$, $(0,\pi)$, (π,π) corresponds to TRIM. Here, since the right-hand side of the equation becomes $+1$ or -1, ν becomes either 0 or 1, namely Z_2 invariant. When $\nu = 1$, the system is 2DTI (quantum spin Hall system) and possesses helical edge state. Equations (1.3) and (1.7) looks completely different with each other but can be understood in a unified way by considering the change in the electronic polarization associated with the time evolution of the Hamiltonian [5].

Expansion to Three-dimensional Systems

Although Chern number can be defined only in a two-dimensional system, we can define the Z_2 index on 3D system. TRIM in three dimensional square lattice are the eight point in the Brillouin zone, $(0,0,0)$, $(\pi,0,0)$, $(0,\pi,0)$, $(0,0,\pi)$, $(0,\pi,\pi)$, $(\pi,0,\pi)$, $(\pi,\pi,0)$, (π,π,π). We can take six 2D planes $x=0, x=\pm\pi, y=0, y=\pm\pi, z=0, z=\pm\pi$ in the 3D Brillouin zone and obtain the Z_2 index exactly in the same way as Eq. (1.7). We call those index as x_0, x_1, y_0, y_1, z_0, and z_1. Note that, these six indices are not independent with each other; the three values, $x_0 x_1, y_0 y_1, z_0 z_1$ are identical by definition. As a result, only four indices out of six is independent. The indices $\nu_0, \nu_1, \nu_2, \nu_3$ are defined as follows:

$$\delta(\Lambda_i) = \frac{\mathrm{Pf}[w(\Lambda_i)]}{\sqrt{\det[w(\Lambda_i)]}}, \tag{1.8}$$

$$(-1)^{\nu_0} = \prod_{n_j=0,\pi} \delta(\Lambda_{n_1,n_2,n_3}), \tag{1.9}$$

$$(-1)^{\nu_i} = \prod_{n_{j\neq i}=0,\pi; n_i=\pi} \delta(\Lambda_{n_1,n_2,n_3}) \ (i=1,2,3). \tag{1.10}$$

Using these four indices, the classification of 3DTI is given as $(\nu_0; \nu_1, \nu_2, \nu_3)$. The most important index of them is ν_0. The material with $\nu_0 = 1$ is called a strong

TI and possesses two-dimensional Dirac surface state in all the planes. On the other hand, the material with $\nu_0 = 0$ and at least one out of ν_1, ν_2, ν_3 is one is called a weak TI, which can be understood as stacking of two dimensional TI. Consequently, the weak TI possess a surface state only in specific planes.

Topological Insulator with Spatial Inversion Symmetry

The actual calculation of Eq. (1.10) is sometimes too complicated to predict the candidate material of topological insulator. For a centrosymmetric material, however, we can largely simplify the $\mathbf{Z}_2$ invariant. The simplification makes it easier to calculate the $\mathbf{Z}_2$ invariant and predict candidate materials of TI from first-principles calculations.

When a crystal possesses the spatial inversion symmetry, we can define the parity of the wavefunction with the space inversion operation Π. We consider the wavefunction $|\psi_\alpha(\Lambda_i)\rangle$ at the TRIM Λ_i.

$$\Pi|\psi_\alpha(\Lambda_i)\rangle = \xi_\alpha(\Lambda_i)|\psi_\alpha(\Lambda_i)\rangle \qquad (1.11)$$

Here, eigenvalue $\xi_\alpha(\Lambda_i) = \pm 1$ gives the parity of the wavefunction.

Using the parity defined above, we can write down the equation as,

$$\delta(\Lambda_i) = \frac{\mathrm{Pf}[w(\Lambda_i)]}{\sqrt{\det[w(\Lambda_i)]}} = \prod_{n=1}^{N} \xi_{2n}(\Lambda_i). \qquad (1.12)$$

Here, note that we take the product of the parity for the one side of the band of Kramers pair for an occupied $2N$ numbers of bands. With this, we can calculate the $\mathbf{Z}_2$ index as follows;

$$(-1)^\nu = \prod_{i=1}^{4 \text{ or } 8} \prod_{n=1}^{N} \xi_{2n}(\Lambda_i). \qquad (1.13)$$

The equation is just the product of the parity eigenvalues of the wavefunctions for all the Kramers pairs n in TRIM Λ_i. As a result, the system becomes topologically nontrivial when -1 appear odd times out of $4N$ numbers of $\xi_{2n}(\Lambda_i)$. The situation is the same for the 3DTI as well.

Different from the former mathematically-complicated definition, this equation helps us to understand the condition for the realization of TI intuitively. For simplicity, we discuss the two-dimensional case. Let me consider the situation where the odd parity band and even parity band are inverted only at one specific TRIM. In this case, the gapless surface Dirac surface appears, so that the surface state connects the two bands, as shown in Fig. 1.2a. On the other hand, when band inversion occurs at the two points out of the four points, a gap can open by a continuous deformation, as shown in Fig. 1.2b. As a result, the gapless surface state is stable (unstable) when band inversion occurs at odd (even) numbers of points.

Fig. 1.2 a The case when band inversion occurs at one point. TRIM cannot open a gap because of the Kramers degeneracy associated with time reversal symmetry. **b** The case when band inversion occurs at two points. A gap can open holding the Kramers degeneracy at TRIM

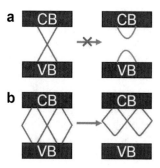

The Edge/Surface State of Topological Insulator

As mentioned earlier, the gapless edge mode appears at the edge/surface of the 2D/3D topological insulator. The Hamiltonian for the edge state of 2DTI can be expressed as follows:

$$H = \hbar v_F \sigma_x k_y, \tag{1.14}$$

where σ is the Pauli matrix of the electron spin. This Hamiltonian means that the spin direction points perpendicular to its momentum, namely spin-momentum locking. Here, the backscattering by the non-magnetic impurity from $+k$ to $-k$ is forbidden at the edge of 2DTI because the electron spin must flip to change the momentum direction. As a result, in an ideal situation, the edge current can flow in a dissipationless manner and result in the quantization of conductance in 2DTI. This is the reason why 2DTI is called a quantum spin Hall system.

Similarly, the surface state Hamiltonian of 3DTI is expressed as follows:

$$H = \hbar v_F (\sigma_x k_y - \sigma_y k_x). \tag{1.15}$$

Here, spin-momentum locking occurs in a two-dimensional surface state. The complete backscattering, which changes the momentum from $+\boldsymbol{k}$ to $-\boldsymbol{k}$, is forbidden in a similar matter to 2DTI. Note that other scattering processes are not forbidden here because of the finite matrix element between the spin components. The exotic band structure of the material makes a unique platform for investigating unconventional transport phenomena and spintronics properties. Moreover, the introduction of ferromagnetism leads to the appearance of the quantum anomalous Hall effect, and the introduction of superconductivity leads to the appearance of topological superconductor and Majorana fermion.

1.1.3   Material Realization of TI

In the former sections, we described the general theory of topological insulator. We here discuss the realization of TI in actual materials. CdTe/HgTe/CdTe [8, 9], AlSb/InAs/GaSb/AlSb [10, 11] quantum well and WTe$_2$ single layer [12–14] are identified as the examples of 2DTI. The existence of the helical edge states and quantum spin Hall effect have been confirmed in these examples [9, 11, 13, 14].

Meanwhile, 3DTI is initially theoretically predicted and experimentally confirmed in $Bi_{1-x}Sb_x$ alloy [7, 15]. This material becomes TI when x is between 0.09 and 0.23. However, in addition to the topological surface state, non-topological surface states cross the Fermi level four times. As a result, this material is not suitable for studying the property of the surface state. As an ideal platform of 3DTI, chalcogenides with tetradymite structure such as Bi_2Se_3, Bi_2Te_3, Sb_2Te_3, $Bi_{2-x}Sb_xTe_{3-y}Se_y$, and $TlBiSe_2$ are proposed [16]. The strained epitaxial HgTe (0.3 %) on top of CdTe is also found to be a 3DTI [7, 17, 18]. Among them, Bi_2Se_3, Bi_2Te_3, and Sb_2Te_3 are the most well studied strong TIs because they possess the simple Dirac cone at the Γ point with a relatively large band gap of about 300 meV. Besides, they are relatively easily cleavage and easy to grow in thin film form because of the layered van der Waals structure [15, 19].

Here, in order to see how the nontrivial band structure of TI emerges, we take the example of Bi_2Se_3 [16].[1] First, we discuss the crystal structure of Bi_2Se_3. The crystal group is rhombohedral $R\bar{3}m$. As shown in Fig. 1.3a Bi_2Se_3 has a layered structure and each layer is composed of the five elements Se-Bi-Se-Bi-Se (quintuple layer, QL). Each layer has a trigonal lattice like Fig. 1.3b and has ABC stacking. Each QL is connected by van der Waals interaction. Since the crystal structure possesses the inversion symmetry, we can easily calculate the Z_2 invariant using the parity of the wavefunction.

We discuss the band structure of the material in the following. In Bi_2Se_3, Bi is Bi^{3+} ([Xe] $4f^{14}5d^{10}6s^2$) and $6p$ band is unoccupied while Se is Se^{2-} ([Ar] $4s^23d^{10}4p^6$). Thus, Bi_2Se_3 is a semiconductor, where the conduction band is composed of Bi $6p$ orbital, and the valence band is composed of Se $4p$ orbital. Next, we consider the bonding and anti-bonding of the orbitals and crystal field splitting and discuss the wavefunction at the Γ point.

First, we consider bonding and anti-bonding. Then, consistent the change of the number of nodes, Se orbital aligns as $P0^-_{x,y,z}$, $P2^+_{x,y,z}$, $P2^-_{x,y,z}$, Bi wavefunction aligns as $P1^+_{x,y,z}$, $P1^-_{x,y,z}$ (Fig. 1.5a(I)). Here the suffix $+$, $-$ means the even or oddness of the parity for a inversion operation. Figure 1.4 shows the structure of the wavefunction. In the following, we consider the crystal field splitting. Reflecting the anisotropy of the crystal structure, each band splits to p_z orbital and p_x, p_y orbitals (Fig.1.5a(II))

Finally, we perturbatively include the spin-orbit interaction. The energy scale of the spin-orbit interaction is smaller than the bonding anti-bonding splitting and crystal field splitting. The spin-orbit interaction mixes the orthogonal bands such as $P1^+_x$, $P1^+_y$, and $P1^+_z$. The spin orbit interaction can be described as $H_{SO} = \lambda l \cdot s = \lambda\{l_z s_z + 1/2(l_z^+ s_z^- + l_z^- s_z^+)\}$. Because of the first term, $l_z s_z$, the energy splitting appears between $P1^+_{x+iy,\uparrow}$, $P1^+_{x-iy,\downarrow}$ and $P1^+_{x+iy,\downarrow}$, $P1^+_{x-iy,\uparrow}$. In addition, because of the second and the third term, $P1^+_{x+iy,\downarrow}$ orbital with $l = 1$, $s = -1/2$ and $P1^+_{z,\uparrow}$ orbital with $l = 0$, $s = 1/2$ mixes and band splitting occurs. As a result, the energy level of

---

[1] Here, we only discuss the case of the Bi_2Se_3, but the same line of discussion apply to Bi_2Te_3 and Sb_2Te_3. Because of the same crystal structure.

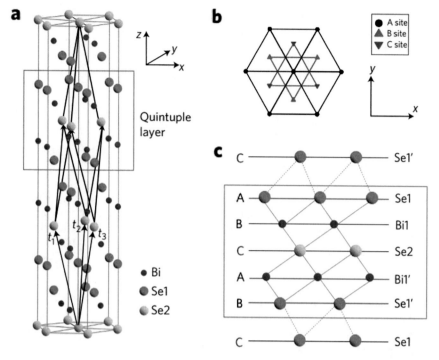

Fig. 1.3 **a** The crystal structure of Bi_2Se_3. **b** The crystal structure seen from the z axis. Each layer is composed of a trigonal lattice. **c** The crystal structure seen from the x axis. Each layer is stacked as A-B-C-A-B-C. Reprinted by permission from Springer Nature: Nature [16], COPYRIGHT (2009)

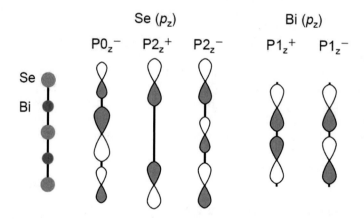

Fig. 1.4 The wavefunctions of Bi_2Se_3. The wavefunctions in the left side have smaller number of nodes and smaller energy

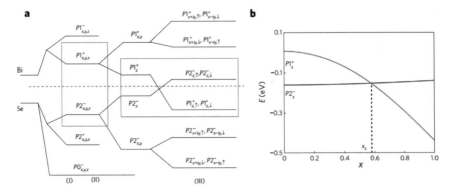

Fig. 1.5 a The energy alignment at Γ point. (I) The bonding anti-bonding orbitals. (II) The crystal field splitting. (III) Spin-orbit interaction. **b** The band inversion of $P1_z^+$ and $P2_z^-$ including the spin-orbit interaction. $x = 0$ corresponds to no spin-orbit interaction. $x = 1$ is the actual size of the spin-orbit interaction. The band inversion occurs at around $x = 0.6$. Reprinted by permission from Springer Nature: Nature [16], COPYRIGHT (2009)

$P1_z^+$ is pushed down. Similarly, the energy level of $P2_z^-$ is pushed up, and the band inversion occurs between these two bands (Fig. 1.5a(III), b). Importantly, the parity of $P2_z^-$ is odd, while the parity of $P1_z^+$ is even. Besides, the band inversion occurs only at around the Γ point. As a result, because of the band inversion, the nearest band to E_F possess $\xi(\Gamma) = +1$ for the Γ point and $\xi(\Lambda) = -1$ for the other TRIM. The occupied band other than the nearest band to the E_F gives the same value of $\xi(\Lambda)$ for all the eight TRIM. As a result, the product $\prod_{i=1}^{8} \xi(\Lambda_i)$ becomes +1 irrespective of their parity. Thus, we get $\prod_{i=1}^{8} \prod_{n=1}^{N} \xi_{2n}(\Lambda_i) = -1$ and $\mathbf{Z}_2$ invariant becomes $\nu_0 = +1$. We can calculate the other index in a similar way, and we get (1; 0, 0, 0). Therefore, it can be concluded that Bi_2Se_3 is a strong 3DTI with a single surface Dirac cone at the Γ point.

To confirm the appearance of topological phase in Bi_2Se_3, we show the result of the first principles calculation. Figure 1.6 is the calculation result of the local density of state for a projected Brillouin zone at the crystal surface. We can clearly see the Dirac dispersion at the surface. Although Sb_2Se_3 is too light (small spin-orbit interaction) to be a TI, Sb_2Te_3 and Bi_2Te_3 becomes TI and the surface state appear as shown in Fig. 1.6.

1.2   Experimental Confirmation of Topological Insulator

As mentioned earlier, since Bi_2Se_3, Bi_2Te_3, and Sb_2Te_3 are among the most well-studied form of a topological insulator, we discuss the experimental confirmation of topological phase in these materials.

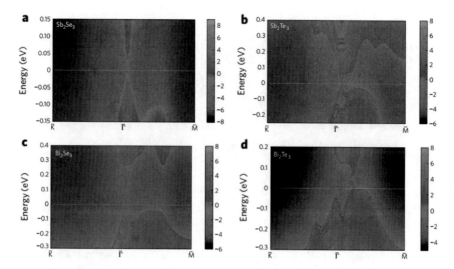

Fig. 1.6 The local density state for a projected Brillouin zone at the crystal surface. **a** Sb_2Se_3. **b** Sb_2Te_3. **c** Bi_2Se_3. **b** Bi_2Te_3. Reprinted by permission from Springer Nature: Nature [16], COPY-RIGHT (2009)

1.2.1 Bulk Crystal

The most sensitive method to study the surface electronic state is Angle-resolved photoemission spectroscopy (ARPES), scanning tunneling microscopy (STM) and scanning tunneling spectroscopy (STS) [15, 19–22]. Since these methods are extremely surface sensitive, they require a clean surface without any impurities. In this context, the layered structures such as Bi_2Se_3, Bi_2Te_3, and Sb_2Te_3 are beneficial because we can get the clean surface by the cleavage of the sample in the high-vacuum chamber.

Angle-Resolved Photoemission Spectroscopy

Figure 1.7 shows the band structure and spin-polarization studied with spin-resolved ARPES [15]. The Dirac dispersion is observed, and the band is shown to be 2D-like from the energy dependence of photon. Moreover, the electron with momentum along the x direction has a spin polarization along y direction, confirming the presence of spin-momentum locking.

Scanning Tunneling Microscopy

By using STM, we can observe the Landau level under magnetic field originating from the topological surface state [21, 22]. In the conventional k^2 dispersion, Landau levels are formed at regular intervals. On the other hand, in the linear Dirac dispersion $E = \hbar v_F k$ like TI and graphene, the position of the Landau level is formed as

$$E_n = v_F \sqrt{2e\hbar |n| B}. \tag{1.16}$$

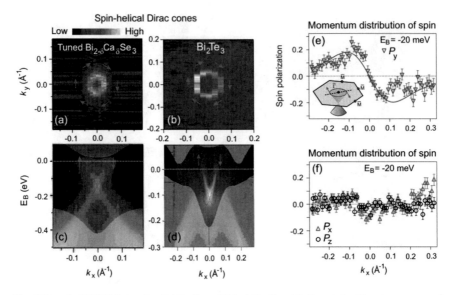

Fig. 1.7 a, b ARPES intensity distribution of $Bi_{2-\delta}Ca_\delta Se_3$, Bi_2Te_3 at the Fermi energy. **c, d** ARPES band dispersion along the cut plane k_x. **e** The y component of the spin polarization. **f** The x component (red) and the y component (black) of the spin polarization. Reprinted by permission from Springer Nature: Nature [15], COPYRIGHT (2009)

Consequently, the energy separation of Landau levels gets closer at higher energy level. In fact, the Landau level is observed under magnetic field as peaks of dI/dV as shown in Fig. 1.8a dI/dV. The peak energy distance gets closer as we go to the higher energy level. As we plot $\sqrt{|n|B}$ at the peak as shown in Fig. 1.8b, we get a single curve and we can see the linear relationship between E_n and $\sqrt{|n|B}$.

These experimental studies reveal the surface state dispersion and spin momentum locking of 3DTI. However, the Fermi level position of these materials are not at the surface and the bulk band instead. Thus, it is difficult to study the transport property of the surface state. Consequently, thin film $(Bi_{1-x}Sb_x)_2Te_3$ is well studied to resolve this problem [23].

1.2.2 Thin Film

In Bi_2Te_3 thin film, n-type carrier is formed because of the Te deficiency and the Fermi level locates at the bulk conduction band. On the other hand, p-type carrier is formed in Sb_2Te_3 thin film due to the antisite defect and the Fermi level locates at the bulk valance band. Here, since the crystal structure of Bi_2Te_3 and Sb_2Te_3 is identical and both of them is a topological insulator. The mixed crystal of the thin film $(Bi_{1-x}Sb_x)_2Te_3$ can compensate the carrier and locate the Fermi level to the surface state. Moreover, the thin film can make the bulk region relatively small and can

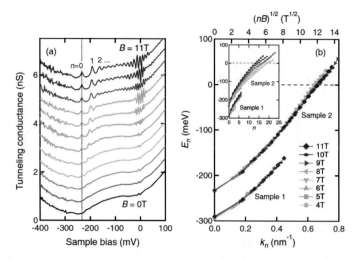

Fig. 1.8 a Tunneling spectroscopy under the magnetic field. The peak position corresponds to the Landau level. **b** The peak position scaled with $\sqrt{|n|B}$. Reprinted figure with permission from [21] Copyright 2010 by the American Physical Society

suppress the parallel conduction from the bulk. Figure 1.9 shows the measurement result by ARPES. By changing the composition x, we can tune the Fermi level to the surface state.

The Surface State Transport Property

Topological insulator exhibit transport property different from the conventional semiconductor because of its surface state. For example, the RT curve shows a insulating behavior at high temperature in bulk $Bi_{2-x}Sb_xTe_{3-y}Se_y$, but their resistance saturates at low temperature because of the surface state [24]. In $(Bi_{1-x}Sb_x)_2Te_3$ thin film, the bulk conduction and the surface transport can be separated by tuning the Fermi level by composition. Figure 1.10 shows the temperature dependence of resistance. Although the end compounds show a metallic behavior, semiconductor behavior appears at the middle composition. This is because the thermally excited carriers in the bulk are deactivated at low temperature.

Figure 1.11 is the magnetic field dependence of Hall resistance. The ordinary Hall effect is enhanced at around $x = 0.94 \sim 0.96$ and it shows a sign reversal from negative at $x = 0.94$ to positive at $x = 0.96$. Since the ordinary Hall effect can be expressed as follows using the carrier density n,

$$R_{yx} = R_0 B, \quad R_0 = \frac{1}{ne}. \tag{1.17}$$

The present result shows that Fermi level goes across the Dirac point in between $x = 0.94$ and $x = 0.96$. The result is consistent with ARPES and shows the effectiveness of Fermi level tuning by composition in $(Bi_{1-x}Sb_x)_2Te_3$ thin film.

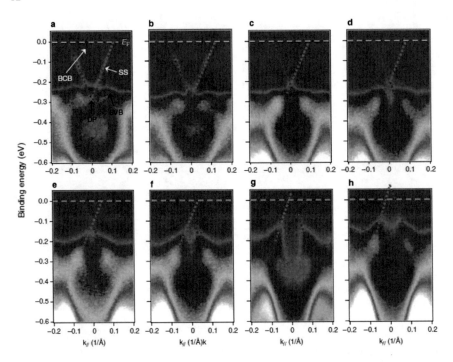

Fig. 1.9 The ARPES result for the $(Bi_{1-x}Sb_x)_2Te_3$ thin film along K-Γ-K direction. The compositions are $x = 0, 0.25, 0.62, 0.75, 0.88, 0.94, 0.96$ and 1.0, respectively. Reprinted bypermission from Springer Nature: Nature [23]., COPYRIGHT (2011)

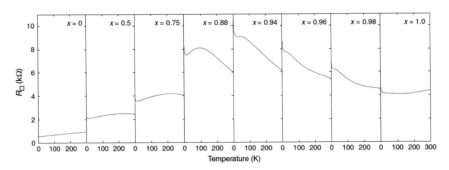

Fig. 1.10 The temperature dependence of resistance of $(Bi_{1-x}Sb_x)_2Te_3$ thin film. Reprinted bypermission from Springer Nature: Nature [23]., COPYRIGHT (2011)

Quantum Hall Effect (QHE)

Since the surface state of TI is two-dimensional, the observation of the quantum Hall effect (QHE) under the magnetic field is expected. In fact, QHE is observed in a TI thin film, $(Bi_{1-x}Sb_x)_2Te_3$ [25]. Since the surface state of TI covers the material, the surface state exists at the top, bottom, and side surfaces. However, since the film

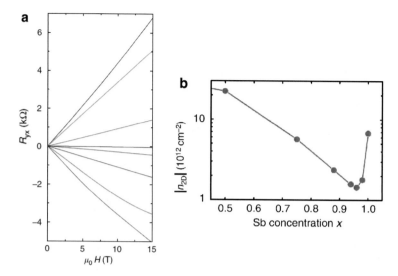

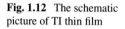

Fig. 1.11 a The magnetic field dependence of Hall resistance. The compositions are x =0.96, 0.98, 1.0, 0, 0.50, 0.75, 0.88, and 0.94, respectively. **b** The composition dependence of 2D sheet carrier density n_{2D}. Reprinted by permission from Springer Nature: Nature [23]., COPYRIGHT (2011)

Fig. 1.12 The schematic picture of TI thin film

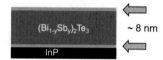

thickness is about 8 nm, the conduction along the side surface is negligible. Thus, we need to take into account the top and bottom surfaces, as shown in Fig. 1.12.

Figure 1.13a, b is the result of the transport measurement. The Hall resistance takes a plateau at around ±25.8 kΩ (h/e^2) and longitudinal resistance goes to zero. Figures 1.13c, d show the conductivity for the sample. The plateau is clearly observed at occupancy $\nu = +1$, and -1, respectively. This can be understood as shown in Fig. 1.13e as the multiple of degeneracy 2 (top and bottom surfaces) to the half-integer QHE of Dirac surface state. Moreover, $\nu = 0$ quantum Hall plateau is observed in a similar way by depositing Sb_2Te_3 as a buffer layer to make the asymmetry between the top and the bottom surfaces [25], Similar QHE of TI has been observed in Bi_2Se_3 on In_2Se_3 buffer layer, exfoliated $BiSbTeSe_2$, or strained HgTe [18, 26–28].

1.3   Experiments on Magnetic Topological Insulator

As discussed in the previous section, the surface state of the topological insulator realizes the exotic band structure, spin-momentum locking, that cannot be realized in a conventional bulk band structure. The peculiar band structure yields various

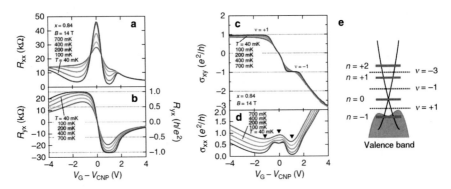

Fig. 1.13 a, b The gate voltage dependence of longitudinal and Hall resistance. **c, d** The gate voltage dependence of longitudinal and Hall conductivity. **e** The relationship between Landau level and Fermi energy in quantum Hall states. Reprinted from [25], Copyright 2015 Macmillan Publishers Limited. Licensed under CC BY 4.0

exotic phenomena through the interaction with magnetism. One such example is the spintronic phenomena, where the spin-momentum locking causes the spin-to-charge conversion and affects the adjacent magnet. Another example is the quantum anomalous Hall effect. The interaction with magnetization opens up a gap at the surface state, and quantization of the anomalous Hall effect occurs when the Fermi level is inside the gap.

One way to study the interaction between magnetism and TI surface state is to attach a ferromagnet to TI. The other way is to dope the magnetic elements to TI, a magnetic topological insulator. Ferromagnetization appears by doping 3d transition metal elements such as Mn, Cr, V into $(Bi_{1-x}Sb_x)_2Te_3$. Here, owing to the layered crystal structure and the large spin-orbit coupling, the perpendicular magnetic anisotropy appears. As a result, Hamiltonian can be expressed as,

$$H = \hbar v_F(\sigma_x k_y - \sigma_y k_x) + m\sigma_z. \tag{1.18}$$

Here, the second term is called a mass term and is proportional to the magnetization. Because of the existence of the mass term, the surface state of a TI will open up a gap. The gap formation of a TI in a magnetic TI has been confirmed through ARPES or quasiparticle interference [29–31]. Figure 1.14 shows the band dispersion of $Cr_x(Bi_{1-y}Sb_y)_{2-x}Te_3$ studied through quasiparticle interference, different from nonmagnetic TI, the band gap of about 50 meV opens at around the Dirac point.

Quantum Anomalous Hall Effect

Quantum anomalous Hall effect (QAHE) is the quantization of the anomalous Hall effect originating from the magnetization. QAHE was theoretically proposed in a band inverted two-dimensional semiconductor or magnetic semiconductor with impurity [33, 34]. With the discovery of topological insulator, the appearance of QAHE is proposed in magnetic proximity coupled topological insulator and magnetic-doped topological insulator [35, 36]. After that the proposal QAHE

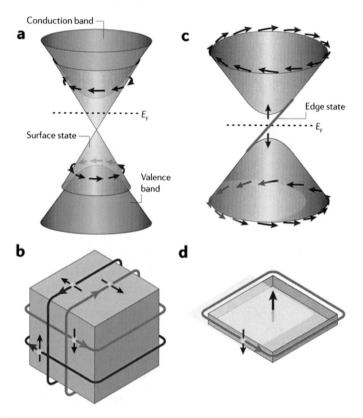

Fig. 1.14 a The massless Dirac-like dispersion of the surface state of a topological insulator. The surface states connects the bulk valence and the bulk conduction bands. **b** Real-space picture of the surface state in a topological insulator. Electrons with spins pointing up and down (red arrows) move in opposite directions due to spin-momentum locking. **c** The massive Dirac dispersion of the surface state in a magnetic topological insulator. **d** The chiral edge mode that appears in a magnetic topological insulator when the Fermi level, E_F, is located in the mass gap induced by the magnetic exchange interaction. The edge electrons conduct electricity without dissipation in one direction along the edge of the sample. Reprinted by permission from Springer Nature: Nature [32], COPYRIGHT (2019)

was experimentally discovered in 3d transition metal element Cr or V doped $(Bi_{1-x}Sb_x)_2Te_3$, $Cr_x(Bi_{1-y}Sb_y)_{2-x}Te_3$ and $V_x(Bi_{1-y}Sb_y)_{2-x}Te_3$ [37–45].

The ferromagnetic moment is induced in $(Bi_{1-x}Sb_x)_2Te_3$ by doping Cr or V. Here, the surface state of magnetic TI will open up a gap, as shown in Fig. 1.14. Quantum anomalous Hall effect appears when we tune the Fermi level inside the gap by Bi, Sb composition tuning, and the application of gate voltage. Figure 1.15 displays the first observation of QAHE. When we tune the gate voltage to $V_G = -1.5$, the Hall resistance is quantized to h/e^2, and longitudinal resistance drops to zero. This means that Hall conductivity is quantized to e^2/h, and longitudinal conductivity becomes zero.

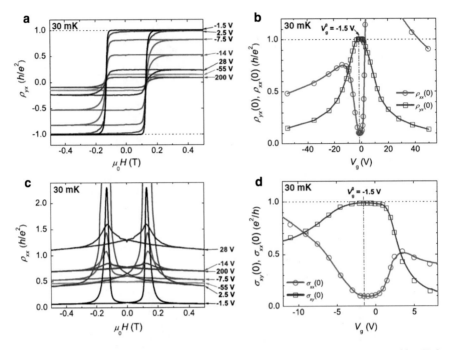

Fig. 1.15 a The magnetic field dependence of Hall resistance at each gate voltage at 30 mK. **b** The gate voltage dependence of longitudinal and Hall resistance. **c** he magnetic field dependence of longitudinal resistance at each gate voltage at 30 mK. **d** The detailed gate voltage dependence of longitudinal and Hall resistance. From [37]. Reprinted with permission from AAAS

This phenomena can be explained as follows. According to Eq. (1.18), the surface state Hamiltonian can be written as

$$H = \boldsymbol{R} \cdot \boldsymbol{\sigma}, \tag{1.19}$$

$$\boldsymbol{R} = (\hbar v_F k_y, -\hbar v_F k_x, \Delta). \tag{1.20}$$

If we calculate the intrinsic contribution of anomalous Hall effect, we get

$$\sigma_{xy} = -\frac{e^2}{h} \int_{BZ} \frac{d\boldsymbol{k}}{4\pi^2} \hat{\boldsymbol{R}} \cdot \left(\frac{\partial \hat{\boldsymbol{R}}}{\partial k_x} \times \frac{\partial \hat{\boldsymbol{R}}}{\partial k_y} \right), \tag{1.21}$$

$$\hat{\boldsymbol{R}} = \frac{\boldsymbol{R}}{R}. \tag{1.22}$$

As a result the calculation gives the half quantized quantum Hall effect as $\sigma_{xy} = \pm \frac{1}{2} \frac{e^2}{h}$. Since topological insulator possesses top and bottom surfaces, these two contributions gives $\sigma_{xy} = \pm \frac{e^2}{h}$.

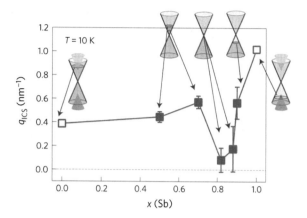

Fig. 1.16 Interface charge to spin conversion efficiency q_{ICS} in $(Bi_{1-x}Sb_x)_2Te_3/Cu/Py$ heterostructure measured with spin-torque FMR as a function of Sb composition. Reprinted by permission from Springer Nature: Nature [52], COPYRIGHT (2016)

QAHE had been realized only in a magnetically-doped topological insulator after the initial discovery in 2013. Quite recently, QAHE is found in two other independent systems; intrinsic antiferromagnetic topological insulator, $MnBi_2Te_4$, and magic-angle twisted bilayer graphene aligned with hBN. $MnBi_2Te_4$ is a layered antiferromagnet with $T_N = 23$ K [46]. Due to the van der Waals nature of the crystal, we can relatively easily exfoliate the crystal down to serveral layers. In odd-layer number $MnBi_2Te_4$ flakes, robust QAHE is observed up to 1.4 K [47]. In addition, Axion insulator to QAHE topological phase transition is induced by the magnetic field in an even-layer number $MnBi_2Te_4$ flakes [48]. In magic-angle twisted bilayer graphene aligned with hBN, flat band is induced by the Moire potential of graphene. Spontaneous orbital magnetization and QAHE is shown to appear in when the band is filled by three-quarter [49, 50]. The realization of new types of QAHE will deepen the understanding of QAHE and lead to a novel functionality associated with the new device geometry.

Spintronics

The interaction between local magnetism and spin-momentum-locked surface conduction electron provides various spintronic functionalities. Magnetic control of the electronic state or electronic control of the magnetization is enabled via charge-to-spin conversion by the Rashba-Edelstein effect [51]: The application of the in-plane electric field shifts the Fermi surface and causes the net spin accumulation perpendicular to the current direction because of the spin-momentum locking at the surface state of TI. Various measurements such as spin-torque ferromagnetic resonance (FMR) and spin-pumping have shown the high efficiency of interface charge-to-spin conversion $q_{ICS} = J_C^{3D}/J_C^{2D}$ (J_C^{3D} is spin current density (Am^{-2}) and J_C^{2D} is surface charge current density). As can be seen from Fig. 1.16, q_{ICS} is about 0.5 nm^{-1} [52], which corresponds to the spin Hall angle θ_{CS} of orders of unity [53–61]. The magnitude of the spin Hall angle is shown to be much larger than the heavy metallic elements such as Pt, β-Ta, β-W, which makes TI an attractive playground to study spintronic phenomena.

1.4 Experiments on Superconducting Proximity Coupled Topological Insulator

Topological surface state possesses spin-momentum locking with no spin degeneracy, namely spinless. The s-wave superconducting proximity effect to such a spinless band induces an exotic p-wave superconductor, which is a topological superconductor. The topological superconductor is getting the keen interest of physicists because it realizes Majorana modes at the sample edge and the vortex core. Majorana particles, which are originally proposed to describe neutrinos, are fermions that are their own antiparticles. Braiding operation of the Majorana fermions in a condensed matter as a composite particle offers topological quantum computation, which is robust against local perturbations [62–64]. For such purpose, various kinds of experiments are conducted to induce superconducting proximity effect to TI surface and observe the Majorana fermion [65–67].

The simplest way to induce the superconducting proximity effect is to attach a superconductor to the surface of TI. For example, Bi_2Se_3 is grown on superconductor $NbSe_2$ and its superconducting proximity effect on the surface to TI has been confirmed with ARPES and STM [68, 69]. In addition, the existence of the zero-bias peak at the vortex core was observed $Bi_2Te_3/NbSe_2$ heterostructure, which was attributed to the formation of Majorana zero mode [70]. The discussion is further supported by the observation of spin selective Andreev reflection [71].

Another interesting experiment is conducted in quantum anomalous Hall state [72]. One experimental approach to form this state has been reported in a junction of superconductor Nb and a magnetic TI, as shown in Fig. 1.17b [72]. Here, TSC is characterized by a Bogoliubov-de Gennes (BdG) Chern number N, which is defined in BdG Hamiltonian with particle-hole symmetry [73]. When the proximitized superconducting gap Δ is larger than the magnetization gap $|m|$, $\Delta > |m|$, TSC state is stabilized and the BdG Chern number becomes $N = \pm 1$. On the other hand, when $\Delta < |m|$, QAH state is realized and the BdG Chern number N corresponds to the Chern number C by $N = 2C = \pm 2$. In this measurement, chiral Majorana edge modes (CMEMs) appear at the interface between the phases with different BdG Chern numbers. Note that one chiral edge mode (CEM) in QAH state can be interpreted as two co-propagating CMEMs, namely one CMEM corresponds to half the CEM. Figure 1.17e shows the magnetic field dependence of Hall conductivity σ_{xy} and the two-terminal conductivity σ_{12} of the sample. Here, the mass gap m is expected to continuously change from minus to plus associated with the magnetization reversal (Fig. 1.17a) [42, 43, 74]. As a result, a sequential topological phase transition from QAH state to TSC state is induced by the magnetic field. Initially, when the magnetization points downward (top left schematic), δ is smaller than $|m|$, so that $C = -1$ and $N = -2$, and the two-terminal conductance σ_{12} is quantized to e^2/h as shown in Fig. 1.17c. When $|m|$ becomes smaller than δ at the magnetization reversal, single CMEMs is formed at the boundary between QAHI ($C = -1$, or $N = -2$) and TSC ($N = -1$). In this situation, one of the two CMEMs is reflected to the current terminal, while the other propagates along the edge of the sample. The splitting leads

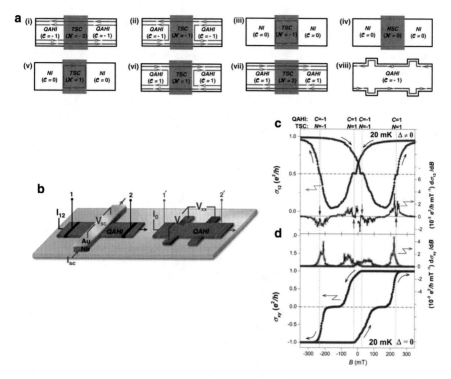

Fig. 1.17 a The evolution of BdG Chern numbers and CMEMs as a function of the magnetic field. **b** The schematic illustration of the measurement procedure for chiral Majorana edge mode (CMEM) at the boundary between QAH state and superconductor. **c** The magnetic field dependence of longitudinal conductivity σ_{12} and Hall conductivity σ_{xy}. Half quantized plateau is observed at the magnetization reversal. From [72]. Reprinted with permission from AAAS

to a half quantization of conductance, $e^2/2h$, which is experimentally observed as a half-quantized plateau [72]. We note that the interpretation of the half-quantized plateau appearing in σ_{12} is still controversial [75–77]. Especially, the realization of high-quality interface between the superconductor and QAHE has shown that half-quantized plateau may be explained by a trivial scenario [78]. The solid confirmation of CMEMs will enable non-Abelian braiding and topological quantum computation in the future [79].

1.5   Purpose of the Thesis

As explained in the above sections, a tremendous amount of experimental study has been conducted on magnetic topological insulators and superconducting proximity coupled topological insulators. Nevertheless, the transport properties and the functionalities have yet to be fully investigated. Here, we focus on the emergent functionalities originating from the spin-momentum locked surface state, mainly focusing on its interaction with magnetism and superconductivity. For this aim, we prepared high-quality thin film with molecular beam epitaxy and investigated its characteristics through transport measurement and magnetic force microscope.

In Chap. 3, we focus on the quantum (anomalous) Hall effect in magnetic topological insulators. Although quantum Hall effect (QHE) and quantum anomalous Hall effect (QAHE) has been realized at the surface state of a topological insulator, the realization temperature of these effects is still below one hundred mK and requires a dilution refrigerator. High-temperature realization of these effects is necessary for the application of a dissipationless channel as well as for expanding the available experimental techniques to obtain further physical insights. To this end, we fabricated a complex heterostructure through the molecular beam epitaxy (MBE) technique. MBE enables us to construct a heterostructure in a relatively easy way. The heterostructure of $(Bi_{1-x}Sb_x)_2Te_3$ and $Cr_x(Bi_{1-y}Sb_y)_{2-x}Te_3$ enabled us to realize a more substantial and homogeneous gap at the surface state. This leads us to realize the higher temperature QHE and QAHE at about $T = 2$ K.

The realization of high-temperature QAHE enabled us to manipulate the chiral edge states further. Because of the discontinuous change in the Chern number, the CES is expected to appear also at the magnetic DW. We design and fabricate the magnetic domains in QAHE with the tip of the magnetic force microscope, and proved the existence of the chiral one-dimensional edge conduction along the prescribed DWs through transport measurement. In addition, we exemplified the proof-of-concept devices based on the reconfigurable CES and Landauer-Büttiker formalism for multiple-domain configurations with the well-defined DW channels.

In Chap. 4, we focus on the spintronic functionalities of magnetic TI. Although topological insulator has been shown to possess a large spin Hall angle, the functionality utilizing these characteristics has not been thoroughly investigated especially in magnetically doped TI. When a spin-momentum locked surface state is coupled to magnetization, the magnetic control of the electronic state and electronic control of magnetization is possible. Magnetically doped TI, $Cr_x(Bi_{1-y}Sb_y)_{2-x}Te_3$ provides one of the ideal platforms to investigate such spintronic functionalities, where strong interaction between conduction electron composed of p-orbital of Bi, Sb, and Te and localized magnetic moment composed of d-orbital of Cr. However, in CBST, the top and bottom surface induces the opposite contribution due to the opposite helicity of spin-momentum locking. Thus, the effects of the top and bottom surfaces will cancel out and the net effect becomes zero. CBST/BST heterostructure provides a solution to this problem because only the top surface interacts with magnetism. Here, we investigate various spintronic functionalities using CBST/BST heterostructures. To study the interaction between local magnetism and surface state, we investigate unidirectional magnetoresistance (UMR) and current nonlinear Hall effect in which resistance value depends on the current direction. In addition, we realize the electronic control of magnetization through current-induced magnetization switching and skyrmion formation and its current-induced motion.

In Chap. 5, we focus on the transport properties of superconducting proximity coupled topological insulator. The superconducting proximity effect on the topological insulator (TI) surface state is one promising way to yield the topological superconductivity. The superconductivity realized at the interface between TI Bi_2Te_3 and non-superconductor FeTe is one of such candidate because the mutual interaction between superconductivity and topological order is expected. In this chapter, to detect the effect of spin-momentum locking in the Cooper pair in Bi_2Te_3/FeTe, we investigate nonreciprocal transport; i.e., current-direction dependent resistance, which is sensitive to the broken inversion symmetry of the electronic state. The nonreciprocal transport is shown to be largely enhanced with superconductivity, meaning that the superconductivity reflects the topological insulator surface state.

Finally, we summarize the results of the thesis in Chap. 6.

References

1. Klitzing KV, Dorda G, Pepper M (1980) New method for high-accuracy determination of the fine-structure constant based on quantized hall resistance. Phys Rev Lett 45(6):494
2. Thouless DJ, Kohmoto M, Nightingale MP, Nijs Md den (1982) Quantized hall conductance in a two-dimensional periodic potential. Phys. Rev. Lett. 49(6):405
3. Kane CL, Mele EJ (2005) Z 2 topological order and the quantum spin hall effect. Phys Rev Lett 95(14):146802
4. Kane CL, Mele EJ (2005) Quantum spin hall effect in graphene. Phys Rev Lett 95(22):226801
5. Fu L, Kane CL (2006) Time reversal polarization and a z 2 adiabatic spin pump. Phys Rev B 74(19):195312
6. Fu L, Kane CL (2007) Topological insulators with inversion symmetry. Phys Rev B 76(4):045302
7. Fu L, Kane CL, Mele EJ (2007) Topological insulators in three dimensions. Phys Rev Lett 98(10):106803
8. Bernevig BA, Hughes TL, Zhang S-C (2006) Quantum spin hall effect and topological phase transition in HgTe quantum wells. Science 314(5806):1757–1761
9. König M, Wiedmann S, Brüne C, Roth A, Buhmann H, Molenkamp LW, Qi X-L, Zhang S-C (2007) Quantum spin hall insulator state in HgTe quantum wells. Science 318(5851):766–770
10. Liu C, Hughes TL, Qi X-L, Wang K, Zhang S-C (2008) Quantum spin hall effect in inverted type-ii semiconductors. Phys Rev Lett 100(23):236601
11. Knez I, Rui-Rui D, Sullivan G (2011) Evidence for helical edge modes in inverted inas/gasb quantum wells. Phys Rev Lett 107(13):136603
12. Qian X, Liu J, Liang F, Li J (2014) Quantum spin hall effect in two-dimensional transition metal dichalcogenides. Science 346(6215):1344–1347
13. Fei Z, Palomaki T, Wu S, Zhao W, Cai X, Sun B, Nguyen P, Finney J, Xu X, Cobden DH (2017) Edge conduction in monolayer wte 2. Nat Phys 13(7):677–682
14. Wu S, Fatemi V, Gibson QD, Watanabe K, Taniguchi T, Cava RJ, Jarillo-Herrero P (2018) Observation of the quantum spin hall effect up to 100 kelvin in a monolayer crystal. Science 359(6371):76–79
15. Hsieh D, Xia Y, Qian D, Wray L, Dil JH, Meier F, Osterwalder J, Patthey L, Checkelsky JG, Ong NP et al (2009) A tunable topological insulator in the spin helical dirac transport regime. Nature 460(7259):1101–1105
16. Zhang H, Liu C-X, Qi X-L, Dai X, Fang Z, Zhang S-C (2009) Topological insulators in bi 2 se 3, bi 2 te 3 and sb 2 te 3 with a single dirac cone on the surface. Nat Phys 5(6):438–442

17. Dai X, Hughes TL, Qi X-L, Fang Z, Zhang S-C (2008) Helical edge and surface states in hgte quantum wells and bulk insulators. Phys Rev B 77(12):125319
18. Brüne C, Liu CX, Novik EG, Hankiewicz EM, Buhmann H, Chen YL, Qi XL, Shen ZX, Zhang SC, Molenkamp LW (2011) Quantum hall effect from the topological surface states of strained bulk hgte. Phys Rev Lett 106(12):126803
19. Xia Y, Qian D, Hsieh D, Wray L, Pal A, Lin H, Bansil A, Grauer DHYS, Hor YS, Cava RJ et al (2009) Observation of a large-gap topological-insulator class with a single dirac cone on the surface. Nat phys 5(6):398–402
20. Roushan P, Seo J, Parker CV, Hor YS, Hsieh D, Qian D, Richardella A, Hasan MZ, Cava RJ, Yazdani A (2009) Topological surface states protected from backscattering by chiral spin texture. Nature 460(7259):1106–1109
21. Hanaguri T, Igarashi K, Kawamura M, Takagi H, Sasagawa T (2010) Momentum-resolved landau-level spectroscopy of dirac surface state in bi 2 se 3. Phys Rev B 82(8):081305
22. Cheng P, Song C, Zhang T, Zhang Y, Wang Y, Jia J-F, Wang J, Wang Y, Zhu B-F, Chen X et al (2010) Landau quantization of topological surface states in bi 2 se 3. Phys Rev Lett 105(7):076801
23. Zhang J, Chang C-Z, Zhang Z, Wen J, Feng X, Li K, Liu M, He K, Wang L, Chen X et al (2011) Band structure engineering in (bi 1–x sb x) 2 te 3 ternary topological insulators. Nat Commun 2(1):1–6
24. Ren Z, Taskin AA, Sasaki S, Segawa K, Ando Y (2011) Optimizing bi 2- x sb x te 3- y se y solid solutions to approach the intrinsic topological insulator regime. Phys Rev B 84(16):165311
25. Yoshimi R, Tsukazaki A, Kozuka Y, Falson J, Takahashi KS, Checkelsky JG, Nagaosa N, Kawasaki M, Tokura Y (2015) Quantum hall effect on top and bottom surface states of topological insulator (bi 1–x sb x) 2 te 3 films. Nat Commun 6(1):1–6
26. Koirala N, Brahlek M, Salehi M, Liang W, Dai J, Waugh J, Nummy T, Han M-G, Moon J, Zhu Y et al (2015) Record surface state mobility and quantum hall effect in topological insulator thin films via interface engineering. Nano Lett 15(12):8245–8249
27. Xu Y, Miotkowski I, Liu C, Tian J, Nam H, Alidoust N, Hu J, Shih C-K, Hasan MZ, Chen YP (2014) Observation of topological surface state quantum hall effect in an intrinsic three-dimensional topological insulator. Nat Phys 10(12):956–963
28. Brüne C, Thienel C, Stuiber M, Böttcher J, Buhmann H, Novik EG, Liu C-X, Hankiewicz EM, Molenkamp LW (2014) Dirac-screening stabilized surface-state transport in a topological insulator. Phys Rev X 4(4):041045
29. Chen YL, Chu J-H, Analytis JG, Liu ZK, Igarashi K, Kuo H-H, Qi XL, Mo S-K, Moore RG, Lu DH et al (2010) Massive dirac fermion on the surface of a magnetically doped topological insulator. Science 329(5992):659–662
30. Xu S-Y, Neupane M, Liu C, Zhang D, Richardella A, Wray LA, Alidoust N, Leandersson M, Balasubramanian T, Sánchez-Barriga J et al (2012) Hedgehog spin texture and berry's phase tuning in a magnetic topological insulator. Nat Phys 8(8):616–622
31. Lee I, Kim CK, Lee J, Billinge SJL, Zhong R, Schneeloch JA, Liu T, Valla T, Tranquada JM, Gu G et al (2015) Imaging dirac-mass disorder from magnetic dopant atoms in the ferromagnetic topological insulator crx (bi0 1sb0 9) 2-xte3. Proc Nat Acad Sci 112(5):1316–1321
32. Tokura Y, Yasuda K, Tsukazaki A (2019) Magnetic topological insulators. Nat Rev Phys 1(2):126–143
33. Onoda M, Nagaosa N (2003) Quantized anomalous hall effect in two-dimensional ferromagnets: quantum hall effect in metals. Phys Rev Lett 90(20):206601
34. Qi X-L, Yong-Shi W, Zhang S-C (2006) Topological quantization of the spin hall effect in two-dimensional paramagnetic semiconductors. Phys Rev B 74(8):085308
35. Qi X-L, Hughes TL, Zhang S-C (2008) Topological field theory of time-reversal invariant insulators. Phys Rev B 78(19):195424
36. Rui Y, Zhang W, Zhang H-J, Zhang S-C, Dai X, Fang Z (2010) Quantized anomalous hall effect in magnetic topological insulators. Science 329(5987):61–64
37. Chang C-Z, Zhang J, Feng X, Shen J, Zhang Z, Guo M, Li K, Yunbo O, Wei P, Wang L-L et al (2013) Experimental observation of the quantum anomalous hall effect in a magnetic topological insulator. Science 340(6129):167–170

38. Checkelsky JG, Yoshimi R, Tsukazaki A, Takahashi KS, Kozuka Y, Falson J, Kawasaki M, Tokura Y (2014) Trajectory of the anomalous hall effect towards the quantized state in a ferromagnetic topological insulator. Nat Phys 10(10):731–736

39. Kou X, Guo S-T, Fan Y, Pan L, Lang M, Jiang Y, Shao Q, Nie T, Murata K, Tang J et al (2014) Scale-invariant quantum anomalous hall effect in magnetic topological insulators beyond the two-dimensional limit. Phys Rev Lett 113(13):137201

40. Kandala A, Richardella A, Kempinger S, Liu C-X, Samarth N (2015) Giant anisotropic magnetoresistance in a quantum anomalous hall insulator. Nat Commun 6(1):1–6

41. Bestwick AJ, Fox EJ, Kou X, Pan L, Wang KL, Goldhaber-Gordon D (2015) Precise quantization of the anomalous hall effect near zero magnetic field. Phys Rev Lett 114(18):187201

42. Kou X, Pan L, Wang J, Fan Y, Choi ES, Lee W-L, Nie T, Murata K, Shao Q, Zhang S-C et al (2015) Metal-to-insulator switching in quantum anomalous hall states. Nat Commun 6(1):1–8

43. Feng Y, Feng X, Yunbo O, Wang J, Liu C, Zhang L, Zhao D, Jiang G, Zhang S-C, He K et al (2015) Observation of the zero hall plateau in a quantum anomalous hall insulator. Phys Rev Lett 115(12):126801

44. Chang C-Z, Zhao W, Kim DY, Zhang H, Assaf BA, Heiman D, Zhang S-C, Liu C, Chan MHW, Moodera JS (2015) High-precision realization of robust quantum anomalous hall state in a hard ferromagnetic topological insulator. Nat Mater 14(5):473–477

45. Chang C-Z, Zhao W, Kim DY, Wei P, Jain JK, Liu C, Chan MHW, Moodera JS (2015) Zero-field dissipationless chiral edge transport and the nature of dissipation in the quantum anomalous hall state. Phys Rev Lett 115(5):057206

46. Otrokov MM, Klimovskikh II, Bentmann H, Estyunin D, Zeugner A, Aliev ZS, Gaß S, Wolter AUB, Koroleva AV, Shikin AM et al (2019) Prediction and observation of an antiferromagnetic topological insulator. Nature 576(7787):416–422

47. Deng Y, Yu Y, Shi MZ, Guo Z, Xu Z, Wang J, Chen XH, Zhang Y (2000) Quantum anomalous hall effect in intrinsic magnetic topological insulator mnbi2te4. Science 367(6480):895–900

48. Liu C, Wang Y, Li H, Wu Y, Li Y, Li J, He K, Xu Y, Zhang J, Wang Y (2020) Robust axion insulator and chern insulator phases in a two-dimensional antiferromagnetic topological insulator. Nat Mater 1–6

49. Sharpe AL, Fox EJ, Barnard AW, Finney J, Watanabe K, Taniguchi T, Kastner MA, Goldhaber-Gordon D (2019) Emergent ferromagnetism near three-quarters filling in twisted bilayer graphene. Science 365(6453):605–608

50. Serlin M, Tschirhart CL, Polshyn H, Zhang Y, Zhu J, Watanabe K, Taniguchi T, Balents L, Young AF (2020) Intrinsic quantized anomalous hall effect in a moiré heterostructure. Science 367(6480):900–903

51. Edelstein VM (1990) Spin polarization of conduction electrons induced by electric current in two-dimensional asymmetric electron systems. Solid State Commun 73(3):233–235

52. Kondou K, Yoshimi R, Tsukazaki A, Fukuma Y, Matsuno J, Takahashi KS, Kawasaki M, Tokura Y, Otani Y (2016) Fermi-level-dependent charge-to-spin current conversion by dirac surface states of topological insulators. Nat Phys 12(11):1027–1031

53. Mellnik AR, Lee JS, Richardella A, Grab JL, Mintun PJ, Fischer MH, Vaezi A, Manchon A, Kim E-A, Samarth N et al (2014) Spin-transfer torque generated by a topological insulator. Nature 511(7510):449–451

54. Wang Y, Deorani P, Banerjee K, Koirala N, Brahlek M, Seongshik O, Yang H (2015) Topological surface states originated spin-orbit torques in bi 2 se 3. Phys Rev Lett 114(25):257202

55. Shiomi Y, Nomura K, Kajiwara Y, Eto K, Novak M, Segawa K, Ando Y, Saitoh E (2014) Spin-electricity conversion induced by spin injection into topological insulators. Phys Rev Lett 113(19):196601

56. Deorani P, Son J, Banerjee K, Koirala N, Brahlek M, Seongshik O, Yang H (2014) Observation of inverse spin hall effect in bismuth selenide. Phys Rev B 90(9):094403

57. Jamali M, Lee JS, Jeong JS, Mahfouzi F, Lv Y, Zhao Z, Nikolić BK, Mkhoyan KA, Samarth N, Wang J-P (2015) Giant spin pumping and inverse spin hall effect in the presence of surface and bulk spin-orbit coupling of topological insulator bi2se3. Nano Lett 15(10):7126–7132

58. Mendes JBS, Santos OA, Holanda J, Loreto RP, De Araujo CIL, Chang C-Z, Moodera JS, Azevedo A, Rezende SM (2017) Dirac-surface-state-dominated spin to charge current conversion in the topological insulator (bi 0.22 sb 0.78) 2 te 3 films at room temperature. Phys Rev B 96(18):180415

59. Wang H, Kally J, Lee JS, Liu T, Chang H, Hickey DR, Mkhoyan KA, Wu M, Richardella A, Samarth N (2016) Surface-state-dominated spin-charge current conversion in topological-insulator–ferromagnetic-insulator heterostructures. Phys Rev Lett 117(7):076601

60. Liu L, Richardella A, Garate I, Zhu Y, Samarth N, Chen C-T (2015) Spin-polarized tunneling study of spin-momentum locking in topological insulators. Phys Rev B 91(23):235437

61. Fan Y, Upadhyaya P, Kou X, Lang M, Takei S, Wang Z, Tang J, He L, Chang L-T, Montazeri M et al (2014) Magnetization switching through giant spin-orbit torque in a magnetically doped topological insulator heterostructure. Nat Mater 13(7):699–704

62. Kitaev A (2006) Anyons in an exactly solved model and beyond. Ann Phys 321(1):2–111

63. Wilczek F (2009) Majorana returns. Nat Phys 5(9):614–618

64. Alicea J (2012) New directions in the pursuit of majorana fermions in solid state systems. Rep Prog phys 75(7):076501

65. Fu L, Kane CL (2008) Superconducting proximity effect and majorana fermions at the surface of a topological insulator. Phys Rev Lett 100(9):096407

66. Hasan MZ, Kane CL (2010) Colloquium: topological insulators. Rev Mod Phys 82(4):3045

67. Qi X-L, Zhang S-C (2011) Topological insulators and superconductors. Rev Mod Phys 83(4):1057

68. Wang M-X, Liu C, Xu J-P, Yang F, Miao L, Yao M-Y, Gao CL, Shen C, Ma X, Chen X et al (2012) The coexistence of superconductivity and topological order in the bi2se3 thin films. Science 336(6077):52–55

69. Su-Yang X, Alidoust N, Belopolski I, Richardella A, Liu C, Neupane M, Bian G, Huang S-H, Sankar R, Fang C et al (2014) Momentum-space imaging of cooper pairing in a half-dirac-gas topological superconductor. Nat Phys 10(12):943–950

70. Xu J-P, Wang M-X, Liu ZL, Ge J-F, Yang X, Liu C, Xu ZA, Guan D, Gao CL, Qian D et al (2015) Experimental detection of a majorana mode in the core of a magnetic vortex inside a topological insulator-superconductor bi 2 te 3/nbse 2 heterostructure. Phys Rev Lett 114(1):017001

71. Sun H-H, Zhang K-W, Lun-Hui H, Li C, Wang G-Y, Ma H-Y, Zhu-An X, Gao C-L, Guan D-D, Li Y-Y et al (2016) Majorana zero mode detected with spin selective andreev reflection in the vortex of a topological superconductor. Phys Rev Lett 116(25):257003

72. He QL, Pan L, Stern AL, Burks EC, Che X, Yin G, Wang J, Lian B, Zhou Q, Choi ES et al (2017) Chiral majorana fermion modes in a quantum anomalous hall insulator–superconductor structure. Science 357(6348):294–299

73. Wang J, Zhou Q, Lian B, Zhang S-C (2015) Chiral topological superconductor and half-integer conductance plateau from quantum anomalous hall plateau transition. Phys Rev B 92(6):064520

74. Wang J, Lian B, Zhang S-C (2014) Universal scaling of the quantum anomalous hall plateau transition. Phys Rev B 89(8):085106

75. Huang Y, Setiawan F, Sau JD (2018) Disorder-induced half-integer quantized conductance plateau in quantum anomalous hall insulator-superconductor structures. Phys Rev B 97(10):100501

76. Ji W, Wen X-G (2018) 1 2 (e 2/h) conductance plateau without 1d chiral majorana fermions. Phys Rev Lett 120(10):107002

77. Lian B, Sun X-Q, Vaezi A, Qi X-L, Zhang S-C (2018) Topological quantum computation based on chiral majorana fermions. Proc Nat Acad Sci 115(43):10938–10942

78. Kayyalha M, Xiao D, Zhang R, Shin J, Jiang J, Wang F, Zhao Y-F, Xiao R, Zhang L, Fijalkowski KM et al (2020) Absence of evidence for chiral majorana modes in quantum anomalous hall-superconductor devices. Science 367(6473):64–67

79. Lian B, Wang J, Sun X-Q, Vaezi A, Zhang S-C (2018) Quantum phase transition of chiral majorana fermions in the presence of disorder. Phys Rev B 97(12):125408

Chapter 2
Experimental Methods

2.1 Thin Film Growth

We grew $(Bi_{1-x}Sb_x)_2Te_3$ and $Cr_x(Bi_{1-y}Sb_y)_{2-x}Te_3$ on InP(111)A substrate and FeTe thin films on CdTe substrate by Molecular Beam Epitaxy (MBE). MBE is a method to grow epitaxial thin film by heating and evaporating the elements (Cr, Bi, Sb, Fe, and Te) in each Knudsen cells (K-cell) as a molecular beam under ultrahigh vacuum. The individual control of the shutter of the K-cell allows the fabrication of heterostructures. Below, we briefly discuss the procedure of film growth.

First, the substrate is clamped to the substrate holder and introduced into the growth chamber through the load-lock chamber after nitrogen blowing. The vacuum of the growth chamber is set to about 10^{-7} Pa so that the molecular beam reaches the substrate without being scattered by the other molecules. Before the growth, we annealed the substrate to obtain a clean surface. The film was grown while rotating the substrate to grow the film homogeneously.

The growth temperature for $(Bi_{1-x}Sb_x)_2Te_3$ and $Cr_x(Bi_{1-y}Sb_y)_{2-x}Te_3$ is about 200 °C. Cr ($x \sim 0.2$) content is estimated from the beam equivalent pressure (BEP) ratio by Cr/(Bi + Sb). Te flux (BEP $= 1.0 \times 10^{-4}$ Pa) was overs-supplied while keeping the Te/(Bi + Sb) ratio at about 40 to suppress the Te vacancies. The growth rate was about 0.2 nm/min. Cr element was modulation-doped by sequentially opening and closing the shutter. After the epitaxial growth of the thin film, annealing under the exposure of Te flux was performed in-situ at 380 °C to get a smoother surface. We capped the film with AlOx with a thickness of about 3 nm at room temperature by an atomic layer deposition system immediately after taking out the samples from the MBE vacuum chamber to suppress the oxidation.

The growth temperatures for FeTe and Bi_2Te_3 were 400 °C and 180 °C, respectively. CdTe(100) substrates were etched with bromine-methanol (bromine 0.01 %) for five minutes before the deposition. For FeTe, the flux ratio was fixed to Fe: Te $= 1:30$ and the growth rate was about 0.7 nm/min. For Bi_2Te_3, the flux ratio was fixed to Bi: Te $= 1:2$ and the growth rate was about 0.2 nm/min.

© Springer Nature Singapore Pte Ltd. 2020
K. Yasuda, *Emergent Transport Properties of Magnetic Topological Insulator Heterostructures*, Springer Theses,
https://doi.org/10.1007/978-981-15-7183-1_2

The grown films were characterized by X-ray diffraction (XRD), atomic force microscope (AFM), and transmission electron microscope (TEM) whenever necessary. The crystal structure, orientation, and the film thickness were studied with XRD. The morphology and uniformity of the film were studied with AFM. The interfacial crystal structure and elemental diffusion were studied with a cross-sectional TEM.

2.2 Device Fabrication

After the growth, we used photolithography to make a field-effect transistor device (Fig. 2.1b), device for the measurement of MFM, and unidirectional magnetoresistance (nonreciprocal measurement) and current-induced magnetization switching.

In device fabrication, we conducted dry etching by Ar ion milling, or wet etching with $H_2O_2:H_3PO_4:H_2O = 1:1:8$ (for $(Bi_{1-x}Sb_x)_2Te_3$ and $Cr_x(Bi_{1-y}Sb_y)_{2-x}Te_3$) and $HCl:H_3P_4:H_2O = 1:1:8$ (for FeTe). We deposited Ti/Au (5/45 nm) using electron-beam (EB) physical vapor deposition system. We used AlO_x with a thickness of about 30 nm as a top gate.

2.3 Electrical Transport Measurement

The transport measurement of the bare sample without device fabrication is conducted with the DC transport option of the Physical Property Measurement System (Quantum Design: PPMS). We used the He3 option of PPMS for a low-temperature measurement below 2 K. A gold wire was attached to a thin film using a silver paste and used as an electrode.

For the device-processed thin film, AC measurement was performed at 13 Hz using lock-in amplifiers (SRS: SR830), as shown in Fig. 2.1b. For current injection, a constant voltage source was used as a constant current of 10 nA by using an attenuator of 100 MΩ.

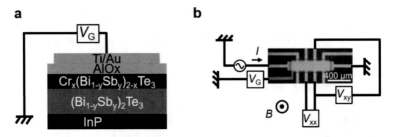

Fig. 2.1 a The cross-sectional view of device. **b** The measurement system of the fabricated device

The second harmonic voltages were measured using a current source (Keithley: Model 6221) and lock-in amplifiers (SRS: SR830). The measurement frequency was fixed to 13 Hz. Most of the measurements were done at 2 K in Physical Property Measurement System (Quantum Design: PPMS) unless otherwise noted.

The pulse current injection with varying pulse heights and the subsequent Hall resistance measurement with enough low current was done with a current source (Keithley: Model 6221) and a voltmeter (Keithley: Model 2182A). The pulse width was set to be ∼1 ms.

2.4 Magnetic Force Microscopy

Magnetic force microscopy (MFM) was conducted under a high vacuum condition with a commercially available scanning probe microscope (attocube AFM/MFM I). The available scan area of the scanner is about 30 μm × 30 μm at $T = 0.5$ K. The piezo block positioner is used for coarse motion. For domain writing with a contact mode, we used the MFMR tip (supplied from NANOSENSORS) with a tip radius of ∼50 nm. The scans were done under a small external magnetic field of 0.015 T ($<B_{c,\mathrm{TI}}$). The scan rate and the step width were set to be 1 μm/s and 50 nm, respectively. For the observation of magnetic contrast with a non-contact mode, we used the PPP-MFMR tip (supplied from NANOSENSORS), which has a smaller tip radius of ∼ 24 nm and has a higher spatial resolution. The measurement was performed at 0.5 K with the cantilever excitation amplitude of 20 nm and the lift height of ∼70 nm. The resonance frequency was 76.202 kHz, and the Q-factor was 6.3×10^4 under the measurement condition. The scan rate and the step width were set to be 1 μm/s and 20 nm, respectively. In situ transport measurement was performed with lock-in amplifiers (SRS: SR830) with the excitation current of 10 nA and 13 Hz frequency. We briefly describe the measurement principle for the magnetic domain structure.

In non-contact mode, we obtain the magnetic domain structure by studying the in-plane distribution of the change of the resonant frequency (f_0) df as shown in Fig. 2.2. The attractive force ($F > 0$) or the repulsive force ($F < 0$) works between the magnetic tip and sample because of the stray field from the magnetic material. The equation of motion of the tip is as follows:

$$m\frac{d^2z}{dt^2} = -kz + F \qquad (2.1)$$

$$= -\left(k - \frac{\partial F}{\partial z}\right)z \qquad (2.2)$$

We make the sample direction seen from the tip as z direction. The resonant frequency of the tip is obtained as $f_0 = 2\pi\sqrt{k/m}$. So the differential along the z direction will

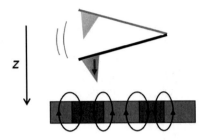

effectively change the spring constant and changes the resonant frequency df as
follows:

$$\frac{df}{f} = -\frac{1}{2k}\frac{\partial F}{\partial z}$$

As a result, df becomes negative (positive) for an attractive (replusive) force.

It is noticeable that the magnetic domain structure itself cannot be observed with
MFM, and it only observes the deviation of the magnetic force made by the stray field
from MFM. In addition, the result cannot be simply understood since the electrostatic
force and van der Waals force works between the tip and the sample.

We used optical interferometry to detect the z position of the tip. The position of
the sample is controlled by a positioner for coarse movement and by a scanner for
fine movement. The scanner for the fine movement is mobile on a scale of 30 µm in
x and y directions.

Chapter 3
Quantum Hall Physics in Magnetic Topological Insulators

3.1 Introduction

Although quantum Hall effect (QHE) and quantum anomalous Hall effect (QAHE) has been realized at the surface state of topological insulator [1–10], the realization temperature of these effects is still below one hundred mK and requires a dilution refrigerator to reach that temperature. High-temperature realization of these effects is necessary for the application of a dissipationless channel as well as for expanding the available experimental techniques to obtain further physical insights. To this end, we fabricated a complex heterostructure through the molecular beam epitaxy (MBE) technique. MBE enables us to construct a heterostructure in a relatively easy way. The heterostructure of $(Bi_{1-x}Sb_x)_2Te_3$ and $Cr_x(Bi_{1-y}Sb_y)_{2-x}Te_3$ enabled us to realize a more substantial and homogenous gap at the surface state. This leads us to realize the higher temperature QHE and QAHE at about $T = 2$ K.

The realization of high-temperature QAHE enabled us to further manipulate the chiral edge states. Because of the discontinuous change in the Chern number, the CES is expected to appear also at the magnetic DW. We design and fabricate the magnetic domains in QAHE with the tip of the magnetic force microscope, and proved the existence of the chiral one-dimensional edge conduction along the prescribed DWs through transport measurement. In addition, we exemplified the proof-of-concept devices based on the reconfigurable CES and Landauer-Büttiker formalism for multiple-domain configurations with the well-defined DW channels.

3.2 Quantum (Anomalous) Hall Effect Stabilized in a Heterostructure

When time-reversal-symmetry is broken with applying enough high magnetic fields or with introducing spontaneous magnetization in 3D-TI, quantum Hall effect (QHE) or quantum anomalous Hall effect (QAHE) emerges as the hallmark of emergent

© Springer Nature Singapore Pte Ltd. 2020
K. Yasuda, *Emergent Transport Properties of Magnetic Topological Insulator Heterostructures*, Springer Theses,
https://doi.org/10.1007/978-981-15-7183-1_3

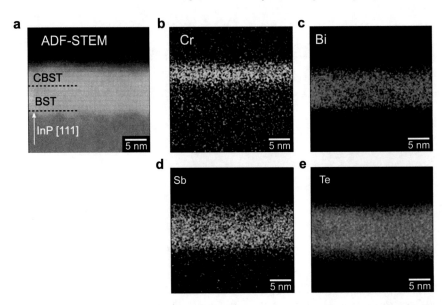

Fig. 3.1 **a** Cross-sectional annular dark-field scanning transmission electron microscopy (ADF-STEM) image of a heterostructure (CBST/BST). **b–e**. Distribution maps of each elements, Cr (**b**), Bi (**c**), Sb (**d**) and Te (**e**) studied by an energy dispersive x-ray spectroscopy (EDX). Cr is distributed only at the top CBST layers with small interdiffusion to the bottom nonmagnetic layers while Bi, Sb and Te are uniformly distributed over the whole layers. Reprinted from [17], Copyright 2015 Macmillan Publishers Limited. Licensed under CC BY 4.0 (Color figure online)

states of 2D electron system [11]. QHE and QAHE are found in 3D-TI thin films [1, 12] or exfoliated thin flakes [13] and their magnetically doped compounds [2–4] Despite the large energy gaps due to the Landau level (LL) formation of the Dirac state (~70 meV at $B = 14$ T) [14] or the presence of the spontaneous magnetization (~50 meV) [15], the observation of QHE or QAHE in thin films has been limited so far at quite low temperatures, typically below 100 mK. This is likely to be because the magnetic impurities and crystalline imperfections make the quantization difficult due to the Landau level broadening.

Here, we present a novel magnetic TI system realizing high-temperature QH effect; semi-magnetic bilayers, namely TI bilayer heterostructures composed of Cr-doped magnetic TI and pristine non-magnetic TI. From a theoretical [16] and experimental [3] points of view, the ground states of QHE and QAHE are unified in the context of topological characterization. In the semi-magnetic TI bilayer, we expect that both magnetization M and magnetic field B drive the Dirac surface states in each TI layer to the QH states, which can be regarded as a hybrid phenomenon of QAHE and QHE with the chiral edge channel. In other words, irrespective of its origin, the system naturally converges to the QH state as lowering the temperature, yielding the quantized Hall conductivity $\sigma_{xy} \sim e^2/h$. Here, the confinement of magnetic ions in a limited region of the heterostructure may suppress the disorder on the whole sample.

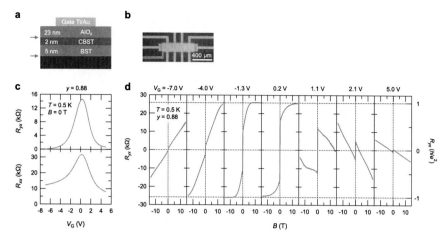

Fig. 3.2 Hall responses R_{yx} in $Cr_x(Bi_{1-y}Sb_y)_{2-x}Te_3/(Bi_{1-y}Sb_y)_2Te_3$ semi-magnetic TI bilayers. **a** CBST and BST represent $Cr_x(Bi_{1-y}Sb_y)_{2-x}Te_3$ and $(Bi_{1-x}Sb_x)_2Te_3$, respectively. The red arrows are the positions where the Dirac state exists. **b** Optical microscope image of a Hall-bar device of CBST (2 nm)/BST (5 nm) ($x \sim 0.2$, $y = 0.88$). **c** Gate voltage V_G dependence of R_{yx} and R_{xx} at $B = 0$ T. **d** Magnetic field dependence of R_{yx} at $T = 0.5$ K for several gate voltage V_G for the device. Reprinted from [17], Copyright 2015 Macmillan Publishers Limited. Licensed under CC BY 4.0

We fabricated TI bilayer heterostructures composed of $Cr_x(Bi_{1-y}Sb_y)_{2-x}Te_3$ (CBST) and $(Bi_{1-x}Sb_x)_2Te_3$ (BST) on semi-insulating InP(111) substrates using molecular-beam epitaxy (See Fig. 3.1 for cross-sectional TEM image), as schematically illustrated in Fig. 3.2a. Here, the surface Dirac states appear on the top surface of CBST as well as at the interface of BST and InP [14], as indicated by red arrows, but not at the interface between two TIs [18, 19]. To tune the E_F by electrical means, we defined a Hall-bar device for $y = 0.88$ with AlO_x gate dielectric and Ti/Au gate electrode. The schematic of the vertical layered structure and top-view photographic image are shown in Fig. 3.2a, b, respectively. R_{yx} and R_{xx} in Fig. 3.2c under the transistor operation at $T = 0.5$ K with $B = 0$ T show systematic behavior as a function of V_G; a single peak is observed at $V_G = 0.2$ V where both R_{yx} and R_{xx} reach the maximum. This means that Fermi energy is right close to the Dirac point. Figure 3.2d is the magnetic field B dependence of R_{yx} measured under respective gate voltage V_G. R_{yx} slope at high B region, originating from the ordinary Hall effect, varies systematically from positive to negative with the change of V_G from negative to positive. The anomalous Hall resistance at $B = 0$ T is enhanced at around $V_G = -1.3$ V and 0.2 V. At these values of gate voltage quantized Hall plateau of R_{yx} reaching $h/e^2 = 25.8$ kΩ is observable with applying a magnetic field.

The quantum Hall effect is also verified even at a higher temperature (e.g. 2 K) for both bare films without gate structure and FET device. Figure 3.3a, b display the magnetic field dependence of Hall and longitudinal conductivity σ_{xy} and σ_{xx} at $T = 2$ K for the bilayers (CBST (2 nm)/BST (5 nm)) of $y = 0.88$ (red) and 0.86

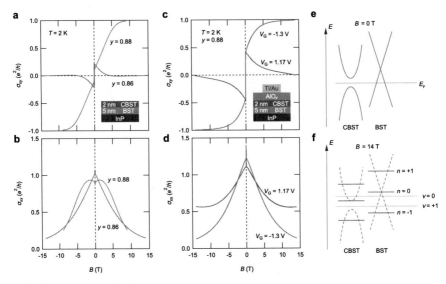

Fig. 3.3 Conductivity responses of semi-magnetic TI bilayers at $T = 2$ K. **a, b** Magnetic field dependence of longitudinal and Hall conductivity σ_{xy} (**a**) and σ_{xx} (**b**) for $y = 0.88$ and $y = 0.86$ bare films of CBST (2 nm)/BST (5 nm) at $T = 2$ K. **c, d** Magnetic field dependence of σ_{xy} (**c**) and σ_{xx} (**d**) for the Hall bar device of $y = 0.88$ at $V_G = -1.3$V ($\nu = +1$) and $V_G = 1.17$ V ($\nu = 0$). **e, f** Schematic band diagram for the surface states of top CBST and bottom BST at $B = 0$ T and $B = 14$ T, respectively. E_F represents the Fermi level at $V_G = 0.2$ V for the $y = 0.88$ device. In (**f**), Landau levels (LLs) $n = +1$, 0 and -1 are denoted by horizontal lines. Filling factor ν for the each location of E_F is displayed. Reprinted from [17], Copyright 2015 Macmillan Publishers Limited. Licensed under CC BY 4.0

(blue) respectively. For the $y = 0.88$ bilayer film, σ_{xy} reaches the quantized value of e^2/h accompanied with the decrease in σ_{xx} towards 0 with increasing B. This is a clear indication of the QH state at the filling factor of $\nu = +1$. In the bilayer device of $y = 0.88$, the similar behavior of $\nu = +1$ QH state is observed at $V_G = -1.3$ V (red) as represented in Fig. 3.3c, d at $B = 14$ T. In contrast, σ_{xy} for the bare bilayer film of $y = 0.86$ and the $y = 0.88$ device at $V_G = 1.17$ V show the asymptotic behavior towards 0 with increasing magnetic field, while σ_{xx} decreases similarly to the case of the $\nu = 1$ QHE (blue). We attributed this to $\nu = 0$ QH state. The contribution of negative ordinary Hall term apparently cancels out the anomalous Hall term, resulting in the $\nu = 0$ state emerging at high B. Noticeably, the QH states in the bilayer films, both $\nu = 0$ and $\nu = +1$, are observable at 2 K, a much higher temperature than that of both QHE in BST [1, 13] and QAHE in CBST films [2–4].

The emergence of two QH states at $\nu = 0$ and $+1$ can be understood from the relationship of the band diagrams of the top and bottom surface bands of BST and CBST shown in Fig. 3.3e, f. Anomalous Hall term, R_{yx} at $B = 0$ T, reaches maximum under $V_G = 0.2$ V, meaning that E_F locates around the center of the gap of CBST surface state (Fig. 3.2e). The ordinary Hall term is added to the anomalous Hall term by applying the magnetic field. The observed positive ordinary Hall term at $V_G = 0.2$

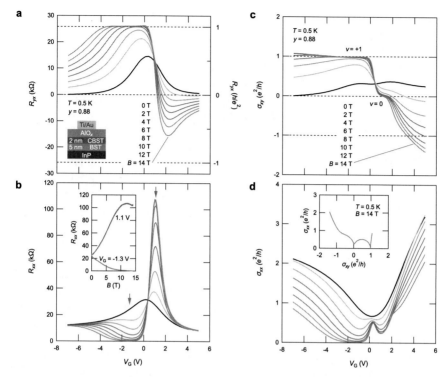

Fig. 3.4 Magnetic field dependence of QH states in gate-tuned semi-magnetic TI bilayers. **a, b,** V_G dependence of R_{yx} (**a**) and R_{xx} (**b**) for $y = 0.88$ FET at $T = 0.5$ K under various magnetic fields. The inset in b shows the magnetic field dependence of R_{xx} at $V_G = 1.1$ V (blue) and $V_G = -1.3$ V (red). Blue and red arrows in the main panel of b represent gate voltages of $V_G = 1.1$ V and -1.3 V, respectively. **c, d** V_G dependence of σ_{xy} (**c**) and σ_{xx} (**d**) at $T = 0.5$ K under various magnetic fields. The inset in d plots the $(\sigma_{xy}(V_G), \sigma_{xx}(V_G))$ at various V_G under magnetic field $B = 14$ T. Reprinted from [17], Copyright 2015 Macmillan Publishers Limited. Licensed under CC BY 4.0 (Color figure online)

V shown in the center panel of Fig. 3.2f indicates that p-type carrier mainly originates from the bottom BST surface since top CBST surface state with E_F within the gap contributes little to the ordinary Hall term. Thus, we can draw the relative energy positions of the Dirac point of CBST and BST surfaces at $V_G = 0.2$ V under $B = 0$ T is as depicted in Fig. 3.3e. Under a high magnetic field, Landau levels (LLs) are formed from the surface Dirac band dispersion as schematically illustrated in Fig. 3.3f. Here, $n = 0$ LL forms on the band edge of the conduction band. $\nu = +1$ QH state emerges when E_F is below $n = 0$ LLs of both top and bottom surfaces. According to the band relationship, as displayed in Fig. 3.3f, fine-tuning of E_F between the two $n = 0$ LLs enables us to achieve $\nu = 0$ state.

To more precisely identify the location of the E_F in the two surface bands drawn in Fig. 3.3e, f, we measured gate voltage dependence of R_{xx} and R_{yx}. The QHE in FET is clearly demonstrated at various magnetic fields in Fig. 3.3a–d. Hallmarks of ν

$= +1$ are observed at around $V_G = -1.3$ V (red arrow in Fig. 3.4b) with increasing B; $R_{yx} \sim 25.8$ kΩ (Fig. 3.4a), $\sigma_{xy} = +e^2/h$ (Fig. 3.4c), and R_{xx} and $\sigma_{xx} \sim 0$ (Figs.3.4b and 3.4d). The other QH plateau at $\nu = 0$ in σ_{xy} (Fig. 3.4c) is realized around $V_G = 1.1$ V; $R_{yx} \sim 0$ (Fig. 3.4a) and high R_{xx} (blue arrow in Fig. 3.4b). Correspondingly, R_{xx} in the inset of Fig. 3.4b largely increases up to about 100 kΩ with increasing B to 14 T. Here, positive anomalous and negative ordinary Hall resistivity almost cancels out, resulting in a small residual R_{yx}. This exemplifies the $\nu = 0$ pseudo-spin Hall insulator state [1], in which we consider the top and bottom degrees of freedom in the surface states as the pseudo spins, as expected from the band diagram shown in Fig. 3.3f. The appearance of $\nu = 0$ in σ_{xy} under high magnetic field between two peaks of σ_{xy} at $B = 0$ T (black curve in Fig. 3.4c) is a supports the discussion based on the band diagram shown above. In fact, the QH states $\nu = 0$ and $+1$ is resolved in the conductivity mapping ($\sigma_{xy}(V_G)$, $\sigma_{xx}(V_G)$) at various V_G, as shown in the inset of the Fig. 3.4d.

In this study, we revealed that the QH state in the bilayers of BST and CBST are observable up to higher temperatures of about 2 K, namely much more stable than those of single-layer films of BST and CBST, which realize QH state typically below 100 mK. This result means that the Cr element in BST inevitably introduces disorder to the system so that the modulation-doping of the Cr element enhances the realization temperature of QH state compared with CBST single layer. This guiding principle is expanded to the QAH state by Mogi et al. [20]. In CBST/BST, only the top surface opens a magnetization gap so that it cannot realize QAH state at zero magnetic field. Instead, by doping the Cr element only close to the top and bottom surfaces, the QAH state is expected to appear. In fact, in the tri-layer and penta-layer structure, as shown in Fig. 3.5b, c, the realization temperature of QAH state is raised. As shown in Fig. 3.5d, e, the Hall resistance is well quantized in a penta-layer structure even at 0.5 K.

3.3   Chiral Edge Conduction on Magnetic Domain Walls

The realization of high-temperature QAHE enables us to further manipulate the chiral edge states (CESs). As mentioned earlier, QAHE is characterized by a Chern number ($C = +1$ or -1), which can be controlled by the magnetization direction ($M > 0$ or $M < 0$). Since the Chern number discontinuously changes at the domain wall (DW) between up and down magnetization, the CES is expected to emerge at the magnetic DW as well (Fig. 3.6a, b) [21, 22]. Similar phenomena is already found in quantum Hall systems, which is characterized by the filling factor of Landau levels in a 2D electron gas under perpendicular magnetic field: The CESs have been experimentally detected not only at the edge of the sample [23, 24] but also at the boundary of different filling factors, namely at the boundary of lateral p-n junction of graphene formed with local electrostatic gating in QH regime [25–27]. In contrast to the CESs at the edge of the sample [2–4, 9, 20, 23, 24, 28] or at the boundary of local gate [25–27], the position of the magnetic DW of QAHE can be manipulated by external stimuli such as local B field, spin-orbit torque by current [29, 30] or heating

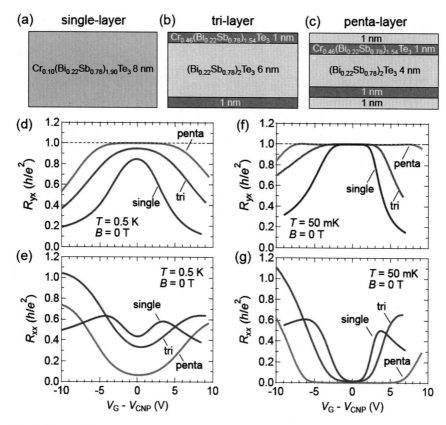

Fig. 3.5 **a–c** Schematic illustration of the sample structures; single-layer CBST 8 nm (**a**), tri-layer CBST(1 nm)/BST(6 nm)/CBST(1 nm) (**b**) and penta-layer BST(1 nm)/CBST(1 nm)/BST(4 nm)/CBST(1 nm)/BST(1 nm). **d–g** Gate voltage V_G dependence of Hall resistance (R_{yx}) and longitudinal resistance (R_{xx}) at $T = 0.5$ K (**d**, **e**) and $T = 50$ mK (**f**, **g**). © 2015 AIP Publishing LLC. Reprinted with permission from [20]

by optical excitation [31]. Therefore, a localized one-way channel based on the CES is constructed in a reconfigurable way. In this work, the quantized conductance of the CES is probed in a controllable manner employing magnetic force microscopy (MFM) in contrast to the previous approaches to probe the CES via naturally-formed DWs [32, 33].

Magnetic TI thin films, Cr-doped $(Bi_{1-x}Sb_x)_2Te_3$, are grown on InP substrates (Fig. 3.6c). QAHE is stabilized by modulation doping of Cr up to about 1 K [20, 28]. Figure 3.6d shows the MFM image taken at 0.5 K in non-contact mode. The naturally-formed maze domain with up (red) and down (blue) M is observed during the magnetization switching (Fig. 3.7) [35]. Instead of using naturally-formed domains, we write the magnetic domain with an arbitrary shape using the stray field from the MFM tip (Fig. 3.8). We first prepare the magnetization directions of the MFM tip and the film to the opposite direction. We then scan over the sample with

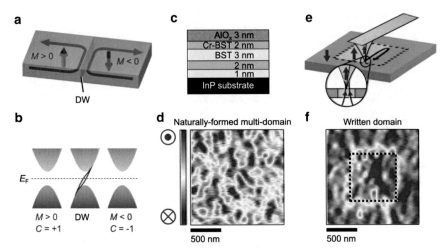

Fig. 3.6 Magnetic domain writing in a magnetic TI by MFM. **a** Illustration of CESs on a magnetic DW. $M < 0$ and $M > 0$ indicate downward (blue) and upward (red) spontaneous magnetization, respectively. **b** The band structures at the domain wall. Exchange interaction opens up the mass gap at the surface state. Two gapless CESs appear due to the discontinuous change of the Chern number from $C = +1$ to -1 at the DW between up and down magnetic domains. **c** Schematic illustration of the Cr modulation-doped magnetic TI. **d** Magnetic domain structure of naturally-formed multi-domain state during the magnetization reversal. The measurement is performed in a non-contact mode. **e** Schematic illustration of the domain writing procedure in a contact mode. The stray field from the MFM tip switches the magnetization of the film, as shown in the enlarged circle. **f** Magnetic domain structure after the domain writing procedure of (**e**). The magnetization is reversed only within the dotted line frame scanned with a contact mode. From [34]. Reprinted with permission from AAAS (Color figure online)

a contact mode inside the dashed frame of Fig. 3.6e under a small magnetic field of 0.015 T (smaller the coercive field of the film $B_{c,TI}$). As can be seen in the MFM image measured with non-contact mode (Fig. 3.6f), the magnetization direction is switched only in the scanned area, meaning that the above procedure is useful for domain writing. We electrically probe the CES employing this writing technique to our Hall-bar devices.

Figure 3.9 shows the magnetic field dependence of the transport properties. Ac current I is applied from contact 5 to contact 6 and voltages V_j are measured at voltage contacts $j = 1 \sim 4$, respectively. In a single-domain states with up or down magnetization (Fig. 3.9a), the Hall resistance is $|R_{13}| = |R_{24}| = 25.1$ kΩ ($\sim h/e^2 = 25.8$ kΩ) with small residual longitudinal resistance of $R_{12} = R_{34} = 1.8$ kΩ at $T = 0.5$ K; this signifies the occurrence of QAHE. The sign of the Hall resistance is reversed at around $B_{c,TI} \sim 0.06$ T, corresponding to the magnetization reversal and hence of the chirality of CES. The longitudinal resistance takes a peak in the multi-domain state around the magnetization reversal. The symmetric resistance values $R_{13} = R_{24}$ and $R_{12} = R_{34}$ in the Hall-bar reflect the small magnetic domain size at the magnetization reversal. [2–4, 9, 20, 28].

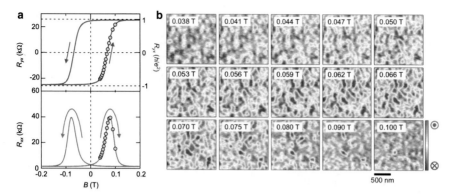

Fig. 3.7 Detailed magnetic field dependence of the domain structure during the magnetization reversal. In the main text, only the MFM image at $B = 0.059$ T is shown. Here, we show the detailed magnetic field dependence of the domain structure at the magnetization reversal. **a** Magnetic field dependence of the Hall resistance R_{yx} and the longitudinal resistance R_{xx}. **b** Magnetic field dependence of the domain structure taken by MFM with a non-contact mode. Gray circles on (**a**) indicate the resistance values of the corresponding MFM images. The red (blue) color corresponds to the up (down) domain. As the magnetic field is applied, small bubble domains start to appear, whose typical size is about 200 nm. The higher magnetic field expands the up domains until the magnetization is completely reversed. From [34]. Reprinted with permission from AAAS (Color figure online)

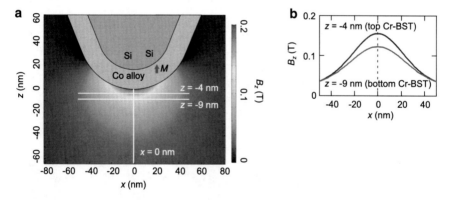

Fig. 3.8 Simulation of the stray field from the tip. To confirm that the magnetization reversal of the sample is caused by the stray field from the MFM tip, we simulate the stray field using the finite element method. The calculation is done under the axisymmetric condition. The magnetization is set upward (z direction). In the MFMR tip (supplied from NANOSENSORS), Co alloy of 17 nm with the magnetization of 300 emu/cc is coated on top of a Si AFM tip and the tip radius is 50 nm. Although the PPP-MFMR tip (supplied from NANOSENSORS) with a smaller radius of ~24 nm is used for the domain observations in Fig. 3.6d, f, the result of the simulation (not shown) is almost the same. **a** Simulation of the z component of the stray field B_z from the tip. **a** The line cut of B_z along x direction at $x = -4$ nm and $x = -9$ nm, corresponding to the center of the two Cr-BST layers, respectively. $B_z > 0.09$ T is applied over a range of 50 nm. The large stray field from the tip enables magnetization reversal of the magnetic TI with a small external magnetic field of 0.015 T. From [34]. Reprinted with permission from AAAS

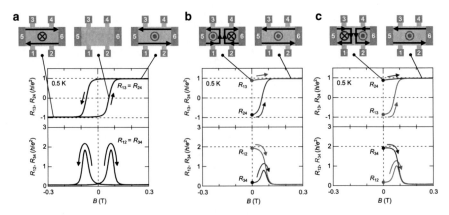

Fig. 3.9 Detection of chiral edge conduction on a magnetic domain wall. **a** Magnetic field dependence of the Hall resistance $R_{13} = R_{24}$ and the longitudinal resistance $R_{12} = R_{34}$ at $T = 0.5$ K. We illustrate corresponding magnetization configurations and CES. Note that the multi-domain state in the upper middle figure is just a schematic view. The actual size of the domain (typically ~ 200 nm) is much smaller than the device size (several tens of μm). **b** Transport properties of the left-up-right-down domain structure and the subsequent magnetic field dependence. **c** Transport properties of the left-down-right-up domain structure and the subsequent magnetic field dependence. From [34]. Reprinted with permission from AAAS

To observe the chiral edge conduction on a single DW, we prepared the left-up-right-down domain structure by writing a magnetic domain with MFM (Fig. 3.9b, upper left schematic). Here, the Hall resistance at the left (right) side R_{13} (R_{24}) is $\sim +h/e^2$ ($\sim -h/e^2$), which corresponds to the magnetization direction. We note that R_{12} and R_{34} are not equal with each other: while R_{12} shows a high resistance of $\sim 2h/e^2$, R_{34} is almost zero. This is in contrast with Fig. 3.9a, where $R_{13} = R_{24}$ and $R_{12} = R_{34}$. When we apply the magnetic field, the resistance goes back to the original value of the single-domain state. This supports that the resistance originates from the left-up-right-down domain structure. The values of $2h/e^2$ and $0h/e^2$ for the upper and lower contacts are explained as follows: two CESs are expected to run in the same direction at the DW of a magnetic TI in QAHE (black arrows in Fig. 3.9b) because of the discontinuous change in Chern number between up and down domains [21, 22]. The two parallel channels intermix with each other and equilibrate across the DW [25–27]. Consequently, the potentials at the downstream of the DW become the same, so that R_{34} is zero. In contrast, R_{12} becomes twice the quantized resistance, $2h/e^2$, since only half of the electrons injected from the current contact 5 is ejected to the other current contact 6. The reversal of the relationship between R_{13} and R_{24} (R_{12} and R_{34}) in the right-up-left-down domain structure (Fig. 3.9c) stems from the inversion of chirality of the CES. Hence, these observations strongly support the existence of the CES at the DW.

We study the DW position dependence of resistance to confirm the existence of the CES at the DW (Fig. 3.10c). The magnetization is reversed gradually from down (blue) to up (red) as the tip scans over the sample from left to right (Fig. 3.10a).

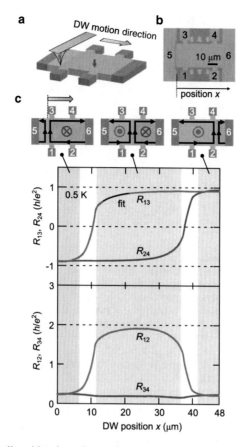

Fig. 3.10 Domain wall position dependence of resistance. **a** Schematics of the measurement procedure. The DW position is continuously pushed from left to right by scanning over the film with an MFM tip under an external magnetic field of 0.015 T. **b** An optical image of the device: modulation-doped magnetic TI (green), gold electrodes (brown). The size of the device is 24 μm × 48 μm (device A). **c** Transport as a function of the DW position x. The areas shaded in color denote the three typical cases of DW position relative to the voltage contacts as schematically displayed in the upper figures. The white shaded areas denote the positions of the contacts. The black line represents the fitting of R_{13} with an exponential function $R_{13} = R_0 - r \exp(-(x - 9\mu m)/\xi)$, where the fitting gives $R_0 = 23$ kΩ, $r = 16$ kΩ and the CES width $\xi = 5.0$ μm. From [34]. Reprinted with permission from AAAS (Color figure online)

Consequently, the DW changes its position gradually; the position of the DW (x) is defined as the distance measured from the edge of the left contact 5, as represented in Fig. 3.10b. With the nearly continuous motion of the DW, the Hall resistance R_{13} (R_{24}) in Fig. 3.10c changes from $\sim -h/e^2$ to $\sim +h/e^2$ at the corresponding voltage contacts. Interestingly, R_{12} has a peak at $\sim 2h/e^2$ when the DW is in between the two contacts. This is consistent with our scenario based on the CED at the DW. When the DW reaches the right-side end, the resistance values recover to the

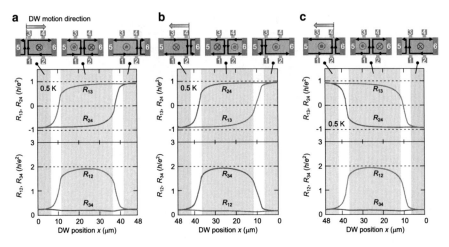

Fig. 3.11 Control experiments of the domain writing. To confirm the validity of the magnetic domain writing and in situ transport measurement, we conduct two control experiments of the DW position dependence experiment, as shown in Fig. 3.10. **a** The DW position x dependence of the transport property scanned from the left (contact 5) to the right (contact 6), which is identical to that in Fig. 3.10a. **b** The DW position x dependence of the transport property scanned from the right (contact 6) to the left (contact 5). The relationship between R_{24} and R_{13} (R_{34} and R_{12}) is reversed due to the chirality inversion of the CES. **c** The DW position x dependence of the transport property scanned from the right (contact 6) to the left (contact 5) (the same as (**b**)), but the magnetization of the MFM tip (the sample) is set to point down (up), opposite to the case of (**b**). The signs of the Hall resistance R_{13} and R_{24} are inverted, and the relationship between R_{12} and R_{34} is reversed as expected. From [34]. Reprinted with permission from AAAS

down magnetization. The DW position dependence is also clearly demonstrated in the control experiments (Fig. 3.11). The width of the CES is also estimated from the width of the resistance transition in Fig. 3.10c. If the CES width were narrow enough, the Hall resistance R_{13} would be a constant value when the DW position is not right on the voltage contacts 1 and 3. However, this does not hold true in our experiment. We attribute this behavior to the finite width of the CES at the DW; if the CES has some spatial distribution, it affects the electrical potential and Hall resistance even when DW is not right on the voltage contacts. Here, we fit a part of the resistance value R_{13} with an exponential function against the distance of the DW measured from the voltage contacts 1 and 3. According to the fitting, the CES width is estimated to be approximately 5 μm, which is much larger than the DW width itself judged from the MFM image in Fig. 3.6f. This value is consistent with the width of the CES estimated from the study of the critical current density of QAHE [36]. This explains the continuous change of resistance at the magnetization reversal, as observed in Fig. 3.9a. The width of the CES is much larger than the typical domain size of ∼200 nm in the multi-domain state (Fig. 3.6d) such that conduction occurs through the hopping between the multiple CESs [7, 8, 37]. This is the reason why

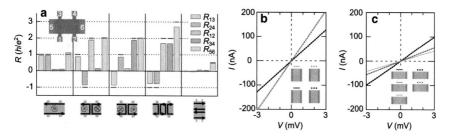

Fig. 3.12 Proof-of-concept demonstration of electronic devices based on CESs. **a** DW configuration dependence of the four-terminal resistance R_{13} (light red), R_{24} (dark red), R_{12} (light blue), R_{34} (dark blue) and the two-terminal resistance R_{56} (light green) under $B = 0$ T. The corresponding domain configurations are displayed under each panel. The resistance is shown in the units of h/e^2. The theoretical calculation based on the Landauer-Büttiker formula are indicated by horizontal solid bars. The measurement for the right end of **a**, and **b** were performed with the Hall-bar with 46 μm in width and 26 μm in length (device **b**) to avoid the interference of CESs. **b** Two-terminal I-V characteristics measured under various DW configurations, where DWs connect the current contacts (brown). The insets show the schematic illustration of the measurement configurations. The solid line and the dotted line almost overlap as expected. **c** The same as (**b**) with DWs connecting the sample edges. From [34]. Reprinted with permission from AAAS (Color figure online)

writing a large magnetic domain is crucial for the observation of the CES at the DW in the present device.

Finally, we show the proof-of-concept experiments of the CES-based electronic devices utilizing the unique CES associated with magnetic DW. We write various domain patterns by the MFM tip and measure their transport properties (Fig. 3.12a). For an ideal CES, we can calculate the resistance using Landauer-Büttiker formalism [38] as shown by the horizontal solid bars. Because of the dissipation-less one-dimensional nature of the CES, the resistance value is determined only by the relative positions of the DWs, independent of the actual size of the device. Although there are some deviations due to the finite size of the Hall-bar and CES, the experimental values present a good agreement with a theoretical calculation. Importantly, in contrast to the cases of QHE [23–27], these resistance states are reconfigurable within a single device. Namely we realize magnetic writing and electrical reading of the domain configuration. Further demonstrations are displayed in Fig. 3.12b, c. This presents a possible device operation with two-terminal resistance measurement. It is noticeable that these two exhibits qualitatively different behaviors depending on the configuration of the DW: when the DWs connect the two current contacts (Fig. 3.12b), the resistance decreases with the number of DWs. On the other hand, when the DWs connect the sample edges (Fig. 3.12c), the resistance increases with the number of DWs. This is consistent with the expectation from the Landauer-Büttiker formalism [38]. The ideal resistance is quantized to $R_{56} = h/(N + 1)e^2$ in the former case and to $R_{56} = (N + 1)h/e^2$ in the latter case, with N being the number of DWs. The present result means that the CES works as a conductive channel when it connects the two current contacts, whereas it works as an edge channel scatterer between the CES at the sample edges when it connects the two sample edges. The qualitative difference stems from the chiral (or one-way) nature of the CES. This is distinct from conductive DWs driven by different physical origins [39–42] e.g. those found in ferroelectric insulators [39, 40] or magnetic insulator with all-in-all-out magnetic order [41, 42].

3.4   Summary

In this section, we discussed the realization of high-temperature QHE and QAHE and the associated control/manipulation of CES at the domain wall.

The realization temperatures of QHE and QAHE was limited to about 100 mK in single layer films of BST and CBST. By making heterostructures of BST and CBST using molecular beam epitaxy (MBE) technique, we realized higher temperature QHE and QAHE up to about $T = 1$ K. The realization of high-temperature QHE and QAHE TI-based heterostructures and superlattices provide a new platform in exploring new functionality and exotic topological phases, such as Weyl semimetal [43, 44].

One such example is the confirmation of CESs at the magnetic DW. We designed and fabricated the magnetic domains on the stable quantum anomalous Hall state with the tip of a magnetic force microscope, and proved the existence of the chiral one-dimensional edge conduction along the prescribed DWs through transport measurements. The proof-of-concept devices based on reconfigurable CESs and Landauer-Büttiker formalism are realized for multiple-domain configurations with well-defined DW channels.

In the future, the experimental methods to control the magnetization configuration in a local, rapid, and reconfigurable manner is required to introduce the CESs into applicable devices. One way to achieve this is to use the spin-orbit torque magnetization switching by the spin accumulation at the surface state [29, 30]. Another way is to use a magneto-optical recording, namely optical writing of magnetization through the modulation of the coercivity by heating [31]. The effective control of the magnetic configuration will lead to the low power-consumption CES-based logic and memory devices [39, 45–47], and quantum information processing [22, 48, 49] in the future.

References

1. Yoshimi R, Tsukazaki A, Kozuka Y, Falson J, Takahashi KS, Checkelsky JG, Nagaosa N, Kawasaki M, Tokura Y (2015) Quantum hall effect on top and bottom surface states of topological insulator (bi 1–x sb x) 2 te 3 films. Nat Commun 6(1):1–6
2. Chang C-Z, Zhang J, Feng X, Shen J, Zhang Z, Guo M, Li K, Yunbo O, Wei P, Wang L-L et al (2013) Experimental observation of the quantum anomalous hall effect in a magnetic topological insulator. Science 340(6129):167–170

3. Checkelsky JG, Yoshimi R, Tsukazaki A, Takahashi KS, Kozuka Y, Falson J, Kawasaki M, Tokura Y (2014) Trajectory of the anomalous hall effect towards the quantized state in a ferromagnetic topological insulator. Nat Phys 10(10):731–736

4. Kou X, Guo S-T, Fan Y, Pan L, Lang M, Jiang Y, Shao Q, Nie T, Murata K, Tang J et al (2014) Scale-invariant quantum anomalous hall effect in magnetic topological insulators beyond the two-dimensional limit. Phys Rev Lett 113(13):137201

5. Kandala A, Richardella A, Kempinger S, Liu C-X, Samarth N (2015) Giant anisotropic magnetoresistance in a quantum anomalous hall insulator. Nat Commun 6(1):1–6

6. Bestwick AJ, Fox EJ, Kou X, Pan L, Wang KL, Goldhaber-Gordon D (2015) Precise quantization of the anomalous hall effect near zero magnetic field. Phys Rev Lett 114(18):187201

7. Kou X, Pan L, Wang J, Fan Y, Choi ES, Lee W-L, Nie T, Murata K, Shao Q, Zhang S-C et al (2015) Metal-to-insulator switching in quantum anomalous hall states. Nat Commun 6(1):1–8

8. Feng Y, Feng X, Yunbo O, Wang J, Liu C, Zhang L, Zhao D, Jiang G, Zhang S-C, He K et al (2015) Observation of the zero hall plateau in a quantum anomalous hall insulator. Phys Rev Lett 115(12):126801

9. Chang C-Z, Zhao W, Kim DY, Zhang H, Assaf BA, Heiman D, Zhang S-C, Liu C, Chan MHW, Moodera JS (2015) High-precision realization of robust quantum anomalous hall state in a hard ferromagnetic topological insulator. Nat Mater 14(5):473–477

10. Chang C-Z, Zhao W, Kim DY, Wei P, Jain JK, Liu C, Chan MHW, Moodera JS (2015) Zero-field dissipationless chiral edge transport and the nature of dissipation in the quantum anomalous hall state. Phys Rev Lett 115(5):057206

11. Rui Y, Zhang W, Zhang H-J, Zhang S-C, Dai X, Fang Z (2010) Quantized anomalous hall effect in magnetic topological insulators. Science 329(5987):61–64

12. Brüne C, Liu CX, Novik EG, Hankiewicz EM, Buhmann H, Chen YL, Qi XL, Shen ZX, Zhang SC, Molenkamp LW (2011) Quantum hall effect from the topological surface states of strained bulk hgte. Phys Rev Lett 106(12):126803

13. Xu Y, Miotkowski I, Liu C, Tian J, Nam H, Alidoust N, Hu J, Shih C-K, Hasan MZ, Chen YP (2014) Observation of topological surface state quantum hall effect in an intrinsic three-dimensional topological insulator. Nat Phys 10(12):956–963

14. Yoshimi R, Tsukazaki A, Kikutake K, Checkelsky JG, Takahashi KS, Kawasaki M, Tokura Y (2014) Dirac electron states formed at the heterointerface between a topological insulator and a conventional semiconductor. Nat Mater 13(3):253–257

15. Chen YL, Chu J-H, Analytis JG, Liu ZK, Igarashi K, Kuo H-H, Qi XL, Mo S-K, Moore RG, Lu DH et al (2010) Massive dirac fermion on the surface of a magnetically doped topological insulator. Science 329(5992):659–662

16. Nagaosa N, Sinova J, Onoda S, MacDonald AH, Ong NP (2010). Anomalous hall effect. Rev Mod Phys 82(2):1539

17. Yoshimi R, Yasuda K, Tsukazaki A, Takahashi KS, Nagaosa N, Kawasaki M, Tokura Y (2015) Quantum hall states stabilized in semi-magnetic bilayers of topological insulators. Nat Commun 6(1):1–6

18. Zhang Y, He K, Chang C-Z, Song C-L, Wang L-L, Chen X, Jia J-F, Fang Z, Dai X, Shan W-Y et al (2010) Crossover of the three-dimensional topological insulator bi 2 se 3 to the two-dimensional limit. Nat Phys 6(8):584–588

19. Zhao Y, Chang C-Z, Jiang Y, DaSilva A, Sun Y, Wang H, Xing Y, Wang Y, He K, Ma X et al (2013) Demonstration of surface transport in a hybrid bi 2 se 3/bi 2 te 3 heterostructure. Sci Rep 3(1):1–7

20. Mogi M, Yoshimi R, Tsukazaki A, Yasuda K, Kozuka Y, Takahashi KS, Kawasaki M, Tokura Y (2015) Magnetic modulation doping in topological insulators toward higher-temperature quantum anomalous hall effect. Appl Phys Lett 107(18):182401

21. Hasan MZ, Kane CL (2010) Colloquium: topological insulators. Rev Mod Phys 82(4):3045

22. Qi X-L, Zhang S-C (2011) Topological insulators and superconductors. Rev Mod Phys 83(4):1057

23. Klitzing KV, Dorda G, Pepper M (1980) New method for high-accuracy determination of the fine-structure constant based on quantized hall resistance. Phys Rev Lett 45(6):494

24. Novoselov KS, Geim AK, Morozov SVB, Jiang D, Katsnelson MI, Grigorieva IVA, Dubonos SVB, Firsov AA (2005) Two-dimensional gas of massless dirac fermions in graphene. Nature 438(7065):197–200
25. Williams JR, DiCarlo L, Marcus CM (2007) Quantum hall effect in a gate-controlled pn junction of graphene. Science 317(5838):638–641
26. Fräßdorf C, Trifunovic L, Bogdanoff N, Brouwer PW (2016) Graphene p n junction in a quantizing magnetic field: conductance at intermediate disorder strength. Phys Rev B 94(19):195439
27. Barbaros Özyilmaz, Pablo Jarillo-Herrero, Dmitri Efetov, Dmitry A Abanin, Leonid S Levitov, and Philip Kim. Electronic transport and quantum hall effect in bipolar graphene p- n- p junctions. *Physical review letters*, 99(16):166804, 2007
28. Mogi M, Kawamura M, Yoshimi R, Tsukazaki A, Kozuka Y, Shirakawa N, Takahashi KS, Kawasaki M, Tokura Y (2017) A magnetic heterostructure of topological insulators as a candidate for an axion insulator. Nat Mater 16(5):516–521
29. Mellnik AR, Lee JS, Richardella A, Grab JL, Mintun PJ, Fischer MH, Vaezi A, Manchon A, Kim E-A, Samarth N et al (2014) Spin-transfer torque generated by a topological insulator. Nature 511(7510):449–451
30. Yasuda K, Tsukazaki A, Yoshimi R, Kondou K, Takahashi KS, Otani Y, Kawasaki M, Tokura Y (2017) Current-nonlinear hall effect and spin-orbit torque magnetization switching in a magnetic topological insulator. Phys Rev Lett 119(13):137204
31. Yeats AL, Mintun PJ, Pan Y, Richardella A, Buckley BB, Samarth N, Awschalom DD (2017) Local optical control of ferromagnetism and chemical potential in a topological insulator. Proc Nat Acad Sci 114(39):10379–10383
32. Checkelsky JG, Ye J, Onose Y, Iwasa Y, Tokura Y (2012) Dirac-fermion-mediated ferromagnetism in a topological insulator. Nat Phys 8(10):729–733
33. Liu M, Wang W, Richardella AR, Kandala A, Li J, Yazdani A, Samarth N, Phuan NP (2016) Large discrete jumps observed in the transition between chern states in a ferromagnetic topological insulator. Sci Adv 2(7):e1600167
34. Yasuda K, Mogi M, Yoshimi R, Tsukazaki A, Takahashi KS, Kawasaki M, Kagawa F, Tokura Y (2017) Quantized chiral edge conduction on domain walls of a magnetic topological insulator. Science 358(6368):1311–1314
35. Malozemoff AP, Slonczewski JC (2016) Magnetic domain walls in bubble materials: advances in materials and device research, vol 1. Academic Press
36. Kawamura M, Yoshimi R, Tsukazaki A, Takahashi KS, Kawasaki M, Tokura Y (2017) Current-driven instability of the quantum anomalous hall effect in ferromagnetic topological insulators. Phys Rev Lett 119(1):016803, 2017
37. Wang J, Lian B, Zhang S-C (2014) Universal scaling of the quantum anomalous hall plateau transition. Phys Rev B 89(8):085106
38. Büttiker M (1988) Absence of backscattering in the quantum hall effect in multiprobe conductors. Phys Rev B 38(14):9375
39. Catalan G, Seidel J, Ramesh R, Scott JF (2012) Domain wall nanoelectronics. Rev Mod Phys 84(1):119
40. Seidel J, Martin LW, He Q, Zhan Q, Chu Y-H, Rother A, Hawkridge ME, Maksymovych P, Yu P, Gajek M et al (2009) Conduction at domain walls in oxide multiferroics. Nat Mater 8(3):229–234
41. Ueda K, Fujioka J, Takahashi Y, Suzuki T, Ishiwata S, Taguchi Y, Kawasaki M, Tokura Y (2014) Anomalous domain-wall conductance in pyrochlore-type nd 2 ir 2 o 7 on the verge of the metal-insulator transition. Phys Rev B 89(7):075127
42. Ma EY, Cui Y-T, Ueda K, Tang S, Chen K, Tamura N, Wu PM, Fujioka J, Tokura Y, Shen Z-X (2015) Mobile metallic domain walls in an all-in-all-out magnetic insulator. Science 350(6260):538–541
43. AA Burkov and Leon Balents (2011) Weyl semimetal in a topological insulator multilayer. Phys Rev Lett 107(12):127205
44. Qi X-L, Hughes TL, Zhang S-C (2008) Topological field theory of time-reversal invariant insulators. Phys Rev B 78(19):195424

45. Parkin SSP, Hayashi M, Thomas L (2008) Magnetic domain-wall racetrack memory. Science 320(5873):190–194
46. Upadhyaya P, Tserkovnyak Y (2016) Domain wall in a quantum anomalous hall insulator as a magnetoelectric piston. Phys Rev B 94(2):020411
47. Tserkovnyak Y, Loss D (2012) Thin-film magnetization dynamics on the surface of a topological insulator. Phys Rev Lett 108(18):187201
48. Fu L, Kane CL (2009) Probing neutral majorana fermion edge modes with charge transport. Phys Rev Lett 102(21):216403
49. Akhmerov AR, Nilsson J, Beenakker CWJ (2009) Electrically detected interferometry of majorana fermions in a topological insulator. Phys Rev Lett 102(21):216404

Chapter 4
Spintronic Phenomena in Magnetic/Nonmagnetic Topological Insulator Heterostructures

4.1 Introduction

Interconversion of angular momentum between conduction electron and local magnetization is one of the central issues of contemporary spintronic research. For example, in normal metal/ferromagnet (NM/FM) heterostructures, accumulated spins via spin current in NM play a key role in manipulating magnetization of FM, which is referred to as spin-orbit torque (SOT) [1, 2]. This enables magnetization switching [2–5] as well as fast domain wall motion [6, 7], which directly leads to computation, logic and memory device applications. In particular, materials with large spin Hall angle θ_{SH} can provide large spin current with minimal Joule heating enabling energy-saving spintronic devices in the future. Among them, topological insulator is attracting keen interest from the basic science as well as from the application point of view. Because of the one-to-one correspondence between charge and spin degrees of freedom, "spin-momentum locking", the current injection into TI induces the spin accumulation at the surface, namely Rashba-Edelstein effect occurs [8]. In fact, TI have been demonstrated to possess quite a large θ_{SH} [9–14] over 1, a few times to two orders of magnitude larger than heavy metal elements such as Pt [3], β-Ta [4] and β-W [5].

When a spin-momentum locked surface state is coupled to magnetization, the magnetic control of the electronic state and electronic control of magnetization is possible. Magnetically doped TI, $Cr_x(Bi_{1-y}Sb_y)_{2-x}Te_3$ provides one of the ideal platforms to investigate such spintronic functionalities, where strong interaction between conduction electron composed of p-orbital of Bi, Sb, and Te and localized magnetic moment composed of d-orbital of Cr. However, in CBST, the top and bottom surfaces induce the opposite contribution because of the opposite helicity of spin-momentum locking. Thus, the effects of the top and bottom surfaces will cancel out, and the net effect becomes zero. CBST/BST heterostructure, which was discussed in the previous chapter, provides a solution to this problem because only the top surface interacts with magnetism. In this chapter, we investigate various spintronic functionalities using CBST/BST heterostructures. We investigate unidirectional magnetore-

© Springer Nature Singapore Pte Ltd. 2020
K. Yasuda, *Emergent Transport Properties of Magnetic Topological Insulator Heterostructures*, Springer Theses,
https://doi.org/10.1007/978-981-15-7183-1_4

47

sistance (UMR) and current nonlinear Hall effect in which resistance value depends on the current direction. In addition, we realize the electronic control of magnetization through current-induced magnetization switching, skyrmion formation, and its current-induced motion.

4.2 Large Unidirectional Magnetoresistance

Transfer and interconservation of angular momentum is at the heart of spintronics [3, 4, 15–17]. The interconversion between various types of quasi-particles, such as electron, photon, phonon, and magnon enables emergent functionalities in solid. One representative example is electrical control of magnetism or vice versa. Current induced magnetization reversal and spin-torque ferromagnetic resonance have been found in various heterostructures based on heavy metal elements (with large spin-orbit coupling) and ferromagnet, such as Pt/Py, Ta/CoFeB and Pt/Co [3, 4, 16, 17]. Here, the spin current generated in heavy metal due to spin Hall effect or Rashab effect is transferred to a ferromagnet, causing magnetization dynamics on a macroscopic scale. Recently, unique diode-like effect, so-called unidirectional (spin Hall) magnetoresistance (UMR), is reported in such heterostructures under in-plane magnetization [18–20]. Here, the resistance value is dependent on both the current J and the magnetization M direction. There, the spin accumulation direction, either parallel or antiparallel with M, at the interface by spin Hall effect is proposed as a major origin of UMR [18–20]. This can be understood in analogy to giant magnetoresistance (GMR) effect [21, 22], which depends on the relationship of magnetizations, parallel or antiparallel, in stacked ferromagnetic layers. In UMR, one of the ferromagnetic layers is replaced with the heavy metal with large spin-orbit coupling. If the spin polarization at the interface governs the amplitude of this effect, the UMR effect is expected to be further enhanced in a topological insulator (TI) due to the spin-momentum locking of the surface state (Fig. 4.1a) [9, 12, 14, 23] .

We investigate UMR of TI heterostructures [9, 24, 25] composed of nonmagnetic TI $(Bi_{1-y}Sb_y)_2Te_3$ (BST) [24, 26] and magnetic TI $Cr_x(Bi_{1-y}Sb_y)_{2-x}Te_3$ (CBST) [27] on insulating InP substrate. We set the nominal compositions of TI heterostructure to be $x \sim 0.2$ and $y \sim 0.86$. We patterned the thin films into the shape of Hall bar of 10 μm in width and 36 μm in length using photolithography and Ar ion milling. We controlled the Fermi energy E_F of the surface state inside the bulk band gap by tuning the composition y, which is confirmed by the Hall effect measurement [24–26, 28]. Thus, top and bottom surfaces with spin momentum locking are the conductive channel [24, 25, 29, 30]. In addition, only one surface adjacent to the Cr-doped layer effectively interacts with magnetism [24, 25]. Thus, from the symmetry consideration of spin-momentum locking (Fig. 4.1a), we can expect that magnetoresistance depends on the relative configuration between surface electron spin and magnetization directions; parallel (Fig. 4.1b) or anti-parallel (Fig. 4.1c).

Figure 4.1e shows the measured magnetoresistance of the heterostructure CBST (3 nm)/BST (5 nm) (Fig. 4.1d). First, we notice that resistance decreases with an

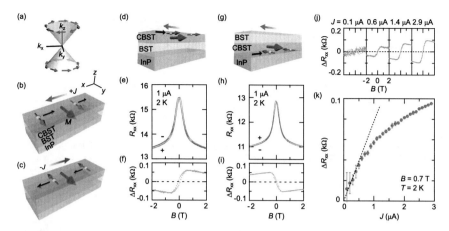

Fig. 4.1 a Schematic diagram of spin-momentum locking of surface Dirac state in TI. **b, c** Schematic illustration of the concept for UMR in TI heterostructures $Cr_x(Bi_{1-y}Sb_y)_{2-x}Te_3/(Bi_{1-y}Sb_y)_2Te_3$ (CBST/BST) on InP substrate under $+J$ (**b**) and $-J$ (**c**) dc current. Here, magnetic field, magnetization and dc current are along the in-plane direction, where dc current is applied perpendicular to the magnetization direction. **d** Schematic illustration of a "normal" CBST/BST heterostructure. **e** Magnetic field dependence of resistance R_{xx} for the sample depicted in (**d**) measured under $J = +1$ μA (red) and $J = -1$ μA (blue) at 2 K. **f** Difference of the resistance ΔR_{xx} of plus and minus current shown in (**e**). **g–i** The same as (**d**)–(**f**) for the "inverted" BST/CBST heterostructure. **j** ΔR_{xx} measured under various current for the normal CBST/BST heterostructure. **k** Current J dependence of ΔR_{xx} at 2 K under $B = 0.7$ T for the normal CBST/BST. The black dotted line shows a slope in the low-J region. Reprinted figure with permission from [31] Copyright 2016 by the American Physical Society (Color figure online)

increasing in-plane magnetic field B. Because of the out-of-plane anisotropy of M in CBST, M points along the z-direction at $B = 0$T, forming the exchange gap in surface Dirac state. As the magnetic field is applied up to 0.7 T, the magnetization direction gradually changes to the in-plane, and therefore, the eventual gap closing of the Dirac surface state causes negative magnetoresistance. Also, we note that the resistance measured under $+1$μA and -1 μA at 2 K show a noticeable deviation, as shown in Fig. 4.1e; the difference in resistance between the two current directions ΔR_{xx} is plotted in Fig. 4.1f. Note that ΔR_{xx} is anti-symmetrized as a function of B and M. ΔR_{xx} is almost zero at 0T where M is pointing out-of-plane, and then increases as the magnetic field increases up to 0.7 T. At higher magnetic field above 0.7 T, ΔR_{xx} becomes almost constant, whose sign is reversed by M reversal in CBST. Furthermore, the inversion of the order of the heterostructure BST (3 nm)/CBST (5 nm) (Fig. 4.1g) reversed sign of ΔR_{xx}, while showing the similar absolute magnitude of UMR, as shown in Fig. 4.1h, i. This is due to the helicity of the spin-momentum locking is opposite between the top and bottom surfaces as depicted in Fig. 4.1d, g. Figure 4.1j, k show the current amplitude dependence of UMR. While ΔR_{xx} shows a negligibly small difference with a current amplitude of 0.1 μA, it gets enhanced with increasing current. The current J dependence of ΔR_{xx} at 0.7 T is summarized in Fig. 4.1k, showing a linear relationship in a low current region, $J < 0.5$ μA. Therefore,

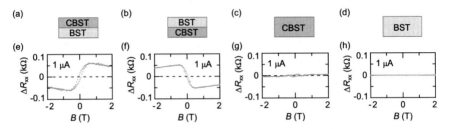

Fig. 4.2 a–d Schematic sample configurations for structure dependent characterization of UMR signal; CBST/BST (**a**), BST/CBST (**b**), CBST (**c**) and BST (**d**). The total thickness of each sample is fixed to 8 nm. **e–h** In-plane magnetic field dependence of ΔR_{xx} at $T = 2$ K. Reprinted figure with permission from [31] Copyright 2016 by the American Physical Society

the relationship between electric field E_x and current density j_x is expressed in a nonlinear form as follows,

$$E_x = R_{xx} j_x + R_{xx}^{(2)} j_x^2. \tag{4.1}$$

Here, $\Delta R_{xx} = 2R_{xx}^{(2)} j_x$ is linearly proportional to current density. The derivation from the linear relationship in Fig. 4.1k at high current ($> 0.5\ \mu$A) is attributed to higher-order effect or heating effect by fairly large current excitation, up to $\Delta T = 2.3$ K at $J = 3\ \mu$A as estimated from the change of R_{xx}. Hereafter, we applied $\pm 1\ \mu$A for the measurements to get enough S/N ratio but to suppress the effect of heating as small as possible.

Figure 4.2 shows the structure dependence of UMR for CBST/BST, BST/CBST, single-layer CBST and single-layer BST. Here, CBST/BST and BST/CBST are taken from Fig. 4.1f, i. In single-layer CBST, ΔR_{xx} takes a small finite value. Here, although the overall shape of the magnetic field dependence is the same as that in CBST/BST, the absolute value of the signal is about ten times smaller. Since Cr is distributed over the whole film in CBST, both the top and bottom surfaces would exhibit finite UMR with an opposite sign, leading to the cancellation of the signal. Here, the observed smaller but finite signal probably originates from the difference in those environments. As for single-layer BST, on the other hand, no UMR signal is observed within the range of measurement error. This reconfirms the magnetic origin of UMR.

Let us compare the magnitude of UMR in the present device with those of previously reported heterostructures in Table 4.1 [18–20]. Since the magnitude of UMR is linear in current, we adopt the quantity $(\Delta R_{xx}/R_{xx})/j$ for a fair comparison. We define the current density by considering each surface conduction thickness of ~ 1 nm in the TI heterostructure [32] (see also the legend of Table 4.1). Even though the current density is much smaller than the other systems, $\Delta R_{xx}/R_{xx}$ is comparable or larger. Therefore, the amplitude of $(\Delta R_{xx}/R_{xx})/j$ is $10^2 - 10^6$ times larger than other bilayer systems, *e.g.* GaMnAs heterostructure or Pt/Co [18–20].

To elucidate a physical origin of such a large UMR, we investigate magnetic-field direction dependence of the signal. Figure 4.3b, c show the in-plane magnetic-field direction dependence (Fig. 4.3a) of normalized ΔR_{xx} and M_y ($\propto \cos\varphi$). $|\Delta R_{xx}|$ is

Table 4.1 Comparison of the magnitude of UMR for various magnetic heterostructures. Note that the current density for CBST/BST are calculated with assuming the thickness of conductive region ~ 2 nm of top and bottom surface states [32]. If the total thickness of the whole film (~ 8 nm) were taken, $\Delta R_{xx}/R_{xx})/j$ is increased by a factor of four. Reprinted figure with permission from [31] Copyright 2016 by the American Physical Society

Material	j (A/cm^2)	R_{xx} (Ω)	ΔR_{xx} (Ω)	$\Delta R_{xx}/R_{xx}$ (%)	$\Delta R_{xx}/R_{xx}/j$ (arb.units)
Ta/Co [19]	10^7	574	0.011	0.0019	1.3
Pt/Co [19]	10^7	176	0.0025	0.0014	1
GaMnAs heterostructure [18]	7.5×10^5	1720	2	0.12	1.1×10^3
CBST/BST (this study)	5.0×10^3	14000	57	0.41	5.7×10^5 *

largest at $B||y$-axis ($\varphi = 0°$, $180°$ and $360°$) and scales well with M_y. Figure 4.3e, f show the out-of-plane magnetic-field direction dependence (Fig. 4.3d) of normalized ΔR_{xx}. Here, M_y and M_z are estimated from the variation of anomalous Hall effect. In accord with the in-plane case, $|\Delta R_{xx}|$ is largest at $B||y$-axis ($\theta = -90°$, $90°$). We note, however, that ΔR_{xx} does not simply scale with M_y. This is perhaps because finite M_z component makes the Dirac dispersion massive, and the resultant hedgehog-like spin texture effectively weakens the spin-momentum locking [33]. To summarize, UMR emerges only when M_y component is finite.

One trivial possibility of such nonlinear magnetoresistance is an additional voltage caused by heat gradient along z-direction such as anomalous Nernst effect and spin Seebeck effect [34–36]. In both processes, induced voltage would be expressed as $V_{thermal} \propto M \times (\nabla T)_z$ [34–36], thus, finite M_y component can cause additional voltage along the x-direction. We can, however, safely exclude this possibility since the additional thermal voltage exhibits the same sign for both heterostructures of CBST/BST/InP (Fig. 4.1d) and BST/CBST/InP (Fig. 4.1g) since InP works as a heat bath; this is inconsistent with the experimentally observed opposite sign of UMR shown in Fig. 4.1f, i. Therefore, the origin of UMR should be explored in intrinsic scattering mechanisms related to electron spins. To clarify the microscopic origin, we studied the temperature dependence under the higher magnetic field (Fig. 4.4a). UMR at low magnetic field decreases with temperature until it almost vanishes at around Curie temperature $T_C \sim 24$ K, supporting its close relevance to ferromagnetism. This is also consistent with the absence of UMR in a single-layer BST film (Fig. 4.1h). Meanwhile, UMR is strongly suppressed at a high magnetic field. This means that it does not simply scale with M_y, and therefore, UMR in TI cannot be explained in terms of the GMR-like mechanism that was proposed in ferromagnet/normal metal bilayers [18–20]. Rather, such a field-induced suppression of $|\Delta R_{xx}|$ is reminiscent of spin Seebeck effect [35, 36] and magnon Hall effect [37], in which the magnon population and hence the signal magnitude are suppressed by gap opening of spin-

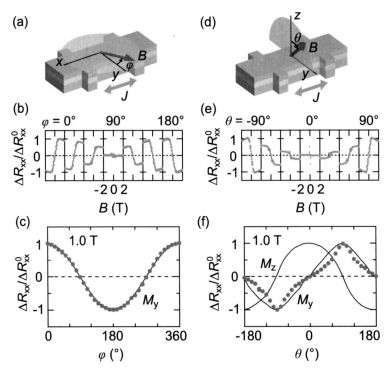

Fig. 4.3 **a** Schematic sample configuration for the measurement of in-plane magnetic field φ dependence of ΔR_{xx}. φ is measured from y-axis. **b** Magnetic field dependence of normalized ΔR_{xx}, $\Delta R_{xx}/\Delta R_{xx}^0$ at respective φ (by 30° step). ΔR_{xx}^0 is defined as ΔR_{xx} at $\varphi = 0°$ and 1.0 T. **c** φ dependence of $\Delta R_{xx}/\Delta R_{xx}^0$ at $B = 1.0 T$. **d–f** The same as (**a**)–(**c**) for the out-of-plane magnetic field θ dependence. θ is measured from z-axis. Here, ΔR_{xx}^0 is ΔR_{xx} at $\theta = 90°$ and 1.0 T. Reprinted figure with permission from [31] Copyright 2016 by the American Physical Society

wave (magnon) at higher field. This leads us to consider the scattering of surface Dirac electrons by magnons as a microscopic origin of UMR.

Conservation of angular momentum leads to the one-way scattering by magnon because of the spin-momentum locking of electron. In magnetic topological insulator, surface state electron and magnetization can be viewed as weakly-coupled systems. For the ease of discussion, we solve surface state electron and dynamics of magnetization independently and consider their interaction as a perturbation. Taking the quantization direction along $M||y$-axis, the angular momentum of magnon is $+1$ (note that spin angular momentum points opposite to magnetization direction). Thus, as shown in Fig. 4.4b, when electron with $s_y = -1/2$ spin (left branch) is back-scattered to $s_y = 1/2$ (right branch), the electron absorbs magnon because of the conservation of angular momentum. Interestingly, although the back-scattering of spin-momentum locked electron is forbidden in a nonmagnetic system, it is allowed in a magnetic system with the aid of angular momentum of magnon. Meanwhile, when electron is scattered from $s_y = 1/2$ to $s_y = -1/2$, it emits magnon as a reverse

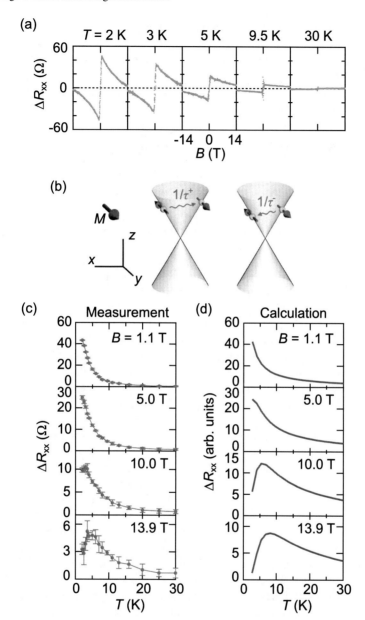

Fig. 4.4 **a** Magnetic field dependence of ΔR_{xx} at various temperatures. **b** Schematic illustration of asymmetric scattering process of spin-polarized surface Dirac electrons by magnon. **c** Temperature dependence of observed ΔR_{xx} under various magnetic fields. **d** Numerical calculation of temperature dependent ΔR_{xx} under various magnetic fields. Reprinted figure with permission from [31] Copyright 2016 by the American Physical Society

process. Phenomena related to such a transfer of angular momentum between electron spin and magnetization is widely recognized in the field of spintronics [15]; for example, spin Seebeck effect [34–36], spin Peltier effect [38], spin pumping [39, 40] and spin torque ferromagnetic resonance [4, 12, 14, 16]. Based on this asymmetric scattering mechanism by magnon, we derive the formula of UMR, employing Boltzmann transport equation under the relaxation time approximation [41] as follows;

$$\Delta R_{xx} \propto j_x \int dk_x \left(-\frac{1}{\tau^+} + \frac{1}{\tau^-} \right) \left(\frac{\partial^2 f}{\partial E^2} \right),$$ (4.2)

$$\frac{1}{\tau^+} \propto \frac{1}{e^{\beta \hbar \omega} - 1} \left(1 - \frac{1}{e^{\beta (\hbar \omega + \hbar v_F k_x - E_F))} + 1} \right),$$ (4.3)

$$\frac{1}{\tau^-} \propto \left(\frac{1}{e^{\beta \hbar \omega} - 1} + 1 \right) \left(1 - \frac{1}{e^{\beta (-\hbar \omega + \hbar v_F k_x - E_F)} + 1} \right).$$ (4.4)

Here, f is the Fermi distribution function and $\hbar \omega$ is the magnon energy with wavenumber $\sim 2k_F$ (k_F : Fermi wavenumber). τ^+ (τ^-) is relaxation time of magnon scattering from left (right) branch to right (left) one. The first factors of Eqs. (4.3) and (4.4) are the probability of magnon absorption and emission, respectively, and the second ones are the probability that the final state of electron is unoccupied. The imbalance between $1/\tau^+$ and $1/\tau^-$ in Eq. (4.2) gives finite UMR in TI. We call such mechanism of UMR as "asymmetric magnon scattering mechanism". Note that Eq. (4.2) is derived for the one-dimensional Dirac dispersion, but the present discussion is readily extended to the actual 2D surface state without essential modification to the scheme. For detailed calculation of UMR, see the supplementary Information of [31].

Figure 4.4c, d show the comparison of the temperature dependence of between the experimental results and calculation based on the above model assuming the magnon energy of $g\mu_B B$ [42] with $g \sim 2$ for the localized Cr moment in CBST. Both results give qualitative consistency; UMR monotonically increases at lower temperatures at a low magnetic field (1.1 T and 5.0 T). At 13.9 T, however, UMR takes a peak structure around the temperature comparable with the magnon gap of around ~ 20 K. Here, the deviation between the experimental result and the numerical calculation above 10 K originates from the breakdown of the spin-wave approximation at temperatures close to $T_C \sim 24$ K. This microscopic model helps us to understand why the UMR in TI is so large: One reason is the spin-momentum locking inherent in TI. Unlike the Rashba interface with two bands having opposite spin helicity, TI with single spin-momentum locking can accumulate spin efficiently without cancellation. Another factor is that TI with tuned E_F around the Dirac point can have a small Fermi momentum k_F lower than $\sim 500 \, \mu m^{-1}$ [24, 26]. Therefore, magnons with small wavenumber and low energy can dominantly contribute to electron scattering, which is easily populated even at low temperatures.

Finally, we discuss the E_F dependence of UMR in the field-effect transistor of TI heterostructure. Here, AlOx layer with a thickness of 30 nm was deposited as a top gate dielectric. Figure 4.5a, b show the gate voltage V_G dependence of R_{yx} and R_{xx}

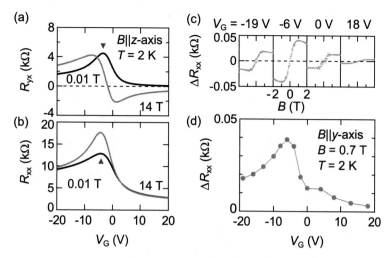

Fig. 4.5 **a, b** Gate voltage V_G dependence of Hall resistance (R_{yx}) and longitudinal resistance (R_{xx}) under magnetic field $B(||z)$ of 0.01 T and 14 T at 2 K. **c, d** Magnetic field and V_G dependence of ΔR_{xx}. The ΔR_{xx} is taken at $B(||y) = 0.7$ T. Reprinted figure with permission from [31] Copyright 2016 by the American Physical Society

under $B||z$-axis at 0.01 T and 14 T. Anomalous Hall effect (R_{yx} measured at 0.01 T, shown by a black line in Fig. 4.5a) and R_{xx} (Fig. 4.5b) take maxima at around -4 V, indicating that E_F of the top surface state is tuned close to the Dirac point [24]. The V_G dependence of UMR measured under $B||y$-axis is summarized in Fig. 4.5c, d. To exclude the V_G dependence of relaxation time, we plot ΔR_{xx}, not $\Delta R_{xx}/R_{xx}$, in Fig. 4.5d. First, the sign of ΔR_{xx} does not change with V_G, *i.e.* irrespective of E_F position in hole ($V_G = -19$ V) and electron ($V_G = 0$ and 18 V) doping regions. This can be understood by considering the scattering process for the hole side in the same way as shown in Fig. 4.4b for the electron side. Moreover, the UMR is maximized at E_F being close to the Dirac point ($V_G = -6$ V). As k_F decreases with E_F approaching the Dirac point, the wavenumber and energy of magnon contributing to the scattering process decrease so that related magnon population increases. This, in combination with the decrease of k_F, results in the maximum ΔR_{xx} and UMR with E_F around the Dirac point.

We remark that, after our report, there is a growing number of evidence that part of the UMR in other material systems such as Pt/Co, Pt/Py can also be explained by the asymmetric magnon scattering mechanisms [43–46]. Thus, the proposed mechanism is not unique to magnetic TI, but applicable to other technologically important materials.

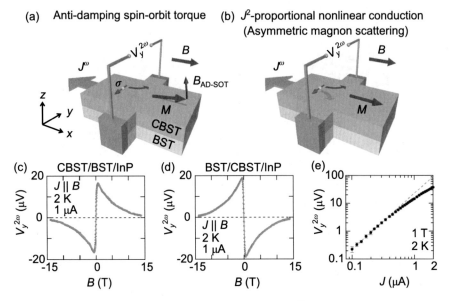

Fig. 4.6 **a** Schematic illustration of 2nd harmonic Hall voltage $V_y^{2\omega}$ caused by anti-damping spin-orbit torque mechanism in CBST/BST. External magnetic field B and ac current J^ω are applied in parallel. Magnetization M is tilted by the effective field due to AD-SOT B_{AD-SOT} caused by the spin accumulation σ. **b** Schematic illustration of $V_y^{2\omega}$ caused by J^2-proportional nonlinear conduction (asymmetric magnon scattering). Nonlinear conduction results in finite $V_y^{2\omega}$ at $M \parallel J$ configuration even though macroscopic magnetization is not tilted. **c** Magnetic field dependence of $V_y^{2\omega}$ at 2 K and 1 µA for the CBST/BST sample. **d** The same as (**c**) for the inverted sample, BST/CBST. **e** Current magnitude J dependence of $V_y^{2\omega}$ at 2 K and 1 T for the CBST/BST sample. The dotted line represents $V_y^{2\omega} \propto J^2$. Reprinted figure with permission from [48] Copyright 2017 by the American Physical Society

4.3 Current-Nonlinear Hall Effect

In the previous section, large current-direction-dependent or unidirectional magnetoresistance (UMR) [1, 18–20, 47] is observed in a magnetic TI under appropriate magnetization and current directions [31]. Such current-nonlinear longitudinal resistance is attributed to the asymmetric scattering of conduction electron by magnon due to the conservation of angular momentum, what we call "asymmetric magnon scattering mechanism" [31].

In this section, we reveal that the asymmetric magnon scattering mechanism causes current-nonlinear resistance also in the transverse direction in magnetic TI under certain configuration, which is observed as the large 2nd harmonic Hall voltage. Importantly, the present result means that θ_{SH} cannot be accurately evaluated by 2nd harmonic technique because the 2nd harmonic voltage is generated not by SOT but mainly by asymmetric magnon scattering. As a target material, we characterize TI heterostructures [24, 25, 31] composed of nonmagnetic TI $(Bi_{1-y}Sb_y)_2Te_3$

(BST) [26, 28] and magnetic TI $Cr_x(Bi_{1-y}Sb_y)_{2-x}Te_3$ (CBST) [49, 50] grown on semi-insulating InP substrate with molecular beam epitaxy (MBE) [24, 25], where the nominal compositions are $x \sim 0.2$ and $y \sim 0.88$. Here, only the top and bottom surfaces are conductive, and the 2nd harmonic voltage is expected only from one surface adjacent to the Cr-doped magnetic layer in a similar way to UMR [25, 31].

Anomalous Hall voltage is typically proportional to current J and out-of-plane component of magnetization M_z. Additional nonlinear transverse voltage proportional to J^2 is allowed from symmetry under $J \parallel M \parallel x$ configuration (Fig. 4.6b). Here, the transverse voltage is expressed as,

$$V_y = R_{AHE} J_x M_z + R_{yx}^{(2)} J_x^2, \tag{4.5}$$

where $R_{yx}^{(2)}$ is a coefficient. Note that there is no ordinary Hall effect and planar Hall effect since they are prohibited in our measurement configuration. The first term, corresponding to anomalous Hall voltage, is zero when M points along the in-plane direction (x). When a large current is applied, however, M is tilted to the out-of-plane direction due to anti-damping SOT (AD-SOT) [1, 2] as shown in Fig. 4.6a such that $M_z = c_{AD-SOT} J_x$, where c_{AD-SOT} is a proportional constant. This is because the effective field by AD-SOT (B_{AD-SOT}) is directed along $\sigma \times M$ [3], where σ is the spin accumulation direction due to the Rashba-Edelstein effect [8]. Thus, under large current and $J \parallel B \parallel x$ configuration, the Hall voltage is expressed as,

$$V_y = R_{AHE} c_{AD-SOT} J_x^2 + R_{yx}^{(2)} J_x^2. \tag{4.6}$$

Both of these two terms are proportional to J^2 (nonlinear against current) but the physical meaning is completely different; macroscopic M is tilted by the AD-SOT mechanism whereas nonlinear conduction occurs even when macroscopic M is unaffected. It is only when the second term of Eq. (4.6) is negligible that we can estimate B_{AD-SOT} from 2nd harmonic Hall voltage [9, 10, 51–53] or current-direction-dependent Hall resistance [3, 4].

To measure 2nd harmonic Hall voltage, we apply ac current by a current source (Keithley: Model 6221) with a frequency of 13 Hz $J^\omega = J \sin(\omega t)$ in the x direction into the Hall bar (10 μm in width and 20 μm in length), then the transverse voltage $V_y = (R_{AHE} c_{AD-SOT} + R_{yx}^{(2)})(\frac{J^2}{2} - \frac{J^2}{2} \cos(2\omega t))$ is measured. Figure 4.6c shows the magnetic field dependence of 2nd harmonic Hall voltage $V_y^{2\omega}$ for the CBST(3 nm)/BST(5 nm) film. At $B = 0$ T, the magnetization of the film points along z direction due to the perpendicular anisotropy [9]. Here, $V_y^{2\omega}$ is almost zero. As the magnetic field is applied up to ~ 0.7 T, M points along x direction. After taking the maximum at ~ 0.7 T, $V_y^{2\omega}$ decreases with an increasing magnetic field. Moreover, as shown in Fig. 4.6d, the sign of $V_y^{2\omega}$ is reversed in the inverted heterostructure, BST(3 nm)/CBST(5 nm)/InP, while showing a comparable magnitude. This is because the helicity of the spin-momentum locking is opposite between the two surface states [31]. Besides, we can safely exclude thermoelectric effects like anomalous Nernst effect or spin Seebeck effect as the primary origin of $V_y^{2\omega}$ in a similar way to the

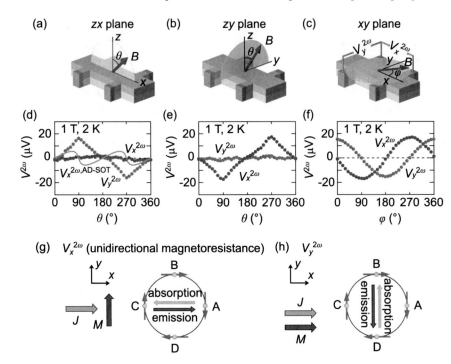

Fig. 4.7 a–c Schematic illustrations of measurement configurations for magnetic-field direction dependence. **d–d** Magnetic-field direction dependence of 2nd harmonic longitudinal voltage $V_x^{2\omega}$ and 2nd harmonic Hall voltage $V_y^{2\omega}$ for zx plane (**d**), zy plane (**e**) and xy plane (**f**). The measurements were done at 1 µA. The green line $V_x^{2\omega,\text{AD–SOT}}$ illustrates the hypothetical 2nd harmonic voltage if $V_y^{2\omega}$ were assumed to be caused purely by the AD-SOT mechanism with magnetization oscillation. **g** Illustration of the origin of $V_x^{2\omega}$ under $J \perp M$ configuration. Red and pink arrows represent the scatterings by magnon emission and absorption processes at around the Fermi surface, respectively. **h** Illustration of the origin of $V_y^{2\omega}$ under $J \parallel M$ configuration. Reprinted figure with permission from [48] Copyright 2017 by the American Physical Society

discussion of UMR above [31]. Figure 4.6e shows the current amplitude dependence of $V_y^{2\omega}$. At low current amplitude, $V_y^{2\omega}$ is proportional to J^2 as expected from Eq. (4.6). The deviation from the proportionality at the larger current region is caused by higher-order effect or heating [31].

Here, we estimate the charge-spin conversion efficiency of the heterostructure assuming that $V_y^{2\omega}$ purely originates from the magnetization oscillation [54]. Here, $V_y^{2\omega}$ is written as

$$V_y^{2\omega} = -\frac{1}{2} \frac{R_{\text{AHE}} B_{\text{AD–SOT}}}{|B| - K} J_x. \tag{4.7}$$

This equation works well when $|B| > K$, where $R_{\text{AHE}} = 2.3\ \text{k}\Omega$ and anisotropy field $K = 0.7\ \text{T}$. Using this, we fit magnetic field dependence of $V_y^{2\omega}$ measured under 1

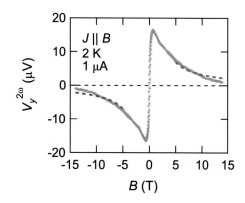

Fig. 4.8 Magnetic field dependence of $V_y^{2\omega}$, which is identical to Fig. 4.6c in the main text. The dotted lines show the fitting curve by Eq. (4.7). Reprinted figure with permission from [48] Copyright 2017 by the American Physical Society

µA condition at B above 5 T as shown in Fig. 4.8. The estimated effective magnetic field amounts to $B_{\text{AD−SOT}} = 26$ mT under 1 µA. Therefore, charge-spin conversion efficiency θ_{CS} is expressed as

$$\theta_{\text{CS}} = j_{\text{S}}/j_{\text{C}} = \frac{2e M_{\text{S}} B_{\text{AD−SOT}} t_{\text{CBST}}}{\hbar j_{\text{C}}} = 160. \qquad (4.8)$$

Here, $t_{\text{CBST}} = 3$ nm, saturation magnetization $M_{\text{S}} = 3.5 \times 10^4$ A/m and $j_{\text{C}} = 1$ µA/(10 µm $\times$ 2 nm) $= 5 \times 10^7$ A/m^2. j_{C} is calculated assuming the conductive regions of top and bottom surfaces to be 1 nm. This value of θ_{CS} is comparable to the previous 2nd harmonic measurements (θ_{SH} :80 - 425) [9, 10] and extraordinarily larger than other estimations (θ_{SH} : 0.4–3.5) [11–14]. Since only the two-dimensional surface state is conductive in TI, it is more appropriate to use the interface charge-spin conversion coefficient $q_{\text{ICS}} = j_{\text{S}}/j_{\text{C}}^{\text{2D}}$ as a measure [14].

$$q_{\text{ICS}} = j_{\text{S}}/j_{\text{C}}^{\text{2D}} = \frac{2e M_{\text{S}} B_{\text{AD−SOT}} t_{\text{CBST}}}{\hbar j_{\text{C}}^{\text{2D}}} = 80 \text{ nm}^{-1}, \qquad (4.9)$$

where $j_{\text{C}}^{\text{2D}} = 1$ µA/10 µm $= 0.1$ A/m. Again, q_{ICS} is two orders of magnitude larger than the previous study of spin-torque ferromagnetic resonance [14]. We argue against the extraordinary value of θ_{SH} derived from the observed large $V_y^{2\omega}$ in the following.

To clarify the origin of the large $V_y^{2\omega}$ signal, we measured magnetic-field direction dependence of 2nd harmonic longitudinal voltage $V_x^{2\omega}$ and 2nd harmonic Hall voltage $V_y^{2\omega}$ in CBST/BST/InP as shown in Fig. 4.7a–f. Let us assume that $V_y^{2\omega}$ originated purely from magnetization oscillation due to AD-SOT. Here, $V_y^{2\omega}$ in Fig. 4.7d should become finite as well, because longitudinal resistance R_{xx} depends on the magnetization direction. Namely R_{xx} is 16.4 kΩ (15.6 kΩ) under $M \parallel z$ ($M \parallel x$). When M oscillates in the zx plane, $V_x^{2\omega}$ should also show a finite value because it modulates longitudinal resistance as well as anomalous Hall resistance. Here, the calculated value from AD-SOT $V_x^{2\omega,\text{AD-SOT}}$ is shown in the green line in Fig. 4.7d as discussed in the following:

When M is oscillating in the zx plane by current-induced AD-SOT, the magnetization direction becomes

$$\theta_M = \theta_M^0 + \frac{\partial \theta_M}{\partial J_x} J_x. \tag{4.10}$$

Here, we define M direction measured from z as θ_M whereas B direction is θ. θ_M^0 is defined as M direction under zero applied current. θ_M dependence of the longitudinal and transverse voltage is described as

$$V_y = R_{\text{AHE}} J_x \cos \theta_M, \tag{4.11}$$

$$V_x = \left(R_{xx}(M \parallel x) + \cos^2 \theta_M \left[R_{xx}(M \parallel z) - R_{xx}(M \parallel x) \right] \right) J_x, \tag{4.12}$$

where $R_{xx}(M \parallel x)$ and $R_{xx}(M \parallel z)$ are the longitudinal resistance at $M \parallel x$ and $M \parallel z$, respectively. Putting Eq. (4.10) into Eqs. (4.11) and (4.12), we get

$$V_y^{2\omega} = -\frac{1}{2} R_{\text{AHE}} J_x^2 \sin \theta_M^0 \frac{\partial \theta_M}{\partial J_x}, \tag{4.13}$$

$$V_x^{2\omega} = -\frac{1}{2} \left[R_{xx}(M \parallel z) - R_{xx}(M \parallel x) \right] J_x^2 \sin 2\theta_M^0 \frac{\partial \theta_M}{\partial J_x}$$
$$= (5.6 \times \sin 2\theta_M^0) \, \mu\text{V}. \tag{4.14}$$

Here, $R_{\text{AHE}} = 2.3 \, \text{k}\Omega$ and $R_{xx}(M \parallel z) - R_{xx}(M \parallel x) = 0.8 \, \text{k}\Omega$ ($R_{xx}(M \parallel z) = 16.4 \, \text{k}\Omega$, $R_{xx}(M \parallel x) = 15.6 \, \text{k}\Omega$) as shown in Fig. 4.9. The second line in Eq. (4.14) is estimated from the experimental value of $V_y^{2\omega} = -\frac{1}{2} R_{\text{AHE}} J_x^2 \frac{\partial \theta_M}{\partial J_x} = 16 \, \mu\text{V}$ at $\theta = \theta_M^0 = 90°$ in Fig. 4.7d. The magnetic field angle θ dependence of the magnetization direction θ_M^0 can be calculated by solving the following equation [54];

$$K \cos \theta_M^0 \sin \theta_M^0 = B \sin(\theta - \theta_M^0). \tag{4.15}$$

Here, anisotropy field $K \sim 0.7 \, \text{T}$ and $B = 1.0 \, \text{T}$, respectively. The theoretical curve $V_x^{2\omega,\text{AD-SOT}}$ in Fig. 4.7d is derived by solving Eqs. (4.14) and (4.15).

One can discuss the contribution of AD-SOT by studying the component of experimental data $V_x^{2\omega}$ in Fig. 4.7d proportional to $\sin 2\theta_M^0$. By fitting $V_x^{2\omega}$ with $C \sin 2\theta_M^0$, the coefficient is $C = 0.1 \pm 0.3 \, \mu\text{V}$. If $V_y^{2\omega}$ was totally originating from AD-SOT contribution, C should become 5.6 μV. Therefore, the contribution of AD-SOT in $V_y^{2\omega}$ can be estimated to be less than $\sim 0.3/5.6 \sim 5$ %.

Therefore, we can conclude that AD-SOT least contributes to $V_x^{2\omega}$ and hence to $V_y^{2\omega}$; namely $V_y^{2\omega}$ should originate mainly from nonlinear conduction (the second term in Eq. (4.6)).

Fig. 4.9 Magnetic field dependence of R_{xx} with $B \parallel x$ at 2 K. At 0 T, M points in the z direction due to the perpendicular anisotropy. At 0.7 T, when M points along the x direction. The mass-gap closing of the Dirac surface state causes negative magnetoresistance [29]. Reprinted figure with permission from [48] Copyright 2017 by the American Physical Society

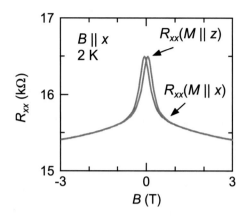

To make further discussions, the microscopic origin of nonlinear conduction has to be clarified. In this context, we notice that $V_y^{2\omega}$ and $V_x^{2\omega}$ have duality; the magnitude of the signal and the angular dependence have a correspondence. Therefore, it is natural to hypothesize that $V_x^{2\omega}$ and $V_y^{2\omega}$ possess the same microscopic origin. In Fig. 4.7d–f, $V_x^{2\omega}$ takes maximum when $M \perp J$. This originates UMR [18–20, 31, 44, 47] as discussed in the previous section [31]. Here, as schematically illustrated in Fig. 4.7g, the J^2-proportional voltage originates from the asymmetry in the relaxation time between the electrons with positive group velocity (position A) and those with negative group velocity (position C) [31]. This is caused by the inequivalence in the scattering mechanism between A and C due to magnon [31, 47]. In a similar way, J^2-proportional transverse voltage ($R_{yx}^{(2)} J_x^2$ in Eqs. (4.5) and (4.6)) can be derived when $M \parallel J$ as shown in Fig. 4.7h. When electrons with positive group velocity (at around position B, angular momentum $+1/2$ with quantization direction along $x \parallel M$) is scattered to around D (angular momentum $-1/2$), the process accompanies the emission of magnon with the angular momentum of $+1$. On the other hand, the scattering of the electron with negative group velocity (at around D, $-1/2$) to around B ($+1/2$) absorbs magnon. This inequivalence in scattering processes leads to the asymmetry in relaxation time. As a result, the electron distribution between around D and around B becomes asymmetric under non-equilibrium conditions, which causes the J^2-proportional transverse term (nonlinear conduction). Such an asymmetric magnon scattering model allows us to evaluate $R_{yx}^{(2)}$. The detailed calculation is described in Supplementary Information of [48]. From that model, we can analytically show that $V_x^{2\omega}(J \parallel x, M \parallel y) = -V_y^{2\omega}(J \parallel x, M \parallel x)/3$; namely $V_x^{2\omega}$ and $V_y^{2\omega}$ show the same order of magnitude, consistent with the experimental results including their signs. The nature of the quantitative deviation of the ratio $V_x^{2\omega}/V_y^{2\omega}$ between the calculation and the experiment is unclear, but we speculate that it stems from the approximations adopted in the calculation [31].

To confirm the validity of the asymmetric magnon scattering model, we numerically calculate the temperature and magnetic field dependence of $V_y^{2\omega}$ and compare it with the experimental results in Fig. 4.10. The experiment and the calculation show

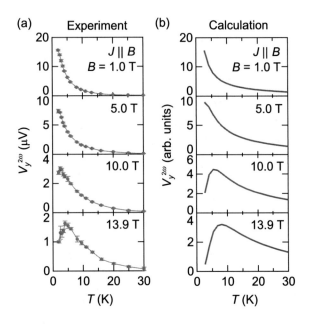

Fig. 4.10 a Temperature dependence of $V_y^{2\omega}$ under various magnetic fields under 1 μA. **b** Calculated results of temperature dependence of $V_y^{2\omega}$. Reprinted figure with permission from [48] Copyright 2017 by the American Physical Society

qualitative consistency. In a similar manner to UMR [31], $V_y^{2\omega}$ shows a monotonic change against temperature at low magnetic fields ($B = 1$ T and 5 T) and almost vanishes at around $T_c \sim 34$ K. Meanwhile, $V_y^{2\omega}$ under a high magnetic field (13.9 T) shows a peak structure. According to the asymmetric magnon scattering model, such a change in the energy scale as a function of the magnetic field B is understood as the variation of magnon gap ($\sim g\mu_B B$). Note that the peak behaviors at high magnetic fields cannot be explained by the AD-SOT-based model since the characteristic energy scale is absent in AD-SOT [2]. Therefore, the results strongly support the asymmetric magnon scattering origin of $V_y^{2\omega}$, namely $R_{AHE}c_{AD-SOT} \ll R_{yx}^{(2)}$. This is the origin of the overestimation of θ_{SH} in magnetic TI [9, 10]. We note that a hysteresis loop shift measurement cannot separate the contribution from asymmetric magnon scattering because the required in-plane magnetic field makes the finite in-plane magnetization and contributes to the Hall effect [55]. Instead, the magneto-optic Kerr effect (MOKE) works as an ideal tool to isolate the contribution of AD-SOT [55].

4.4 Current-Induced Magnetization Switching

In the previous section, we have revealed that AD-SOT makes only a tiny contribution to the to the nonlinear Hall effect. Then, the natural question to ask is, in what current region AD-SOT shows up in our magnetic TI. We note that the above argument on nonlinear Hall effect applies not only to the ac resistance measurement

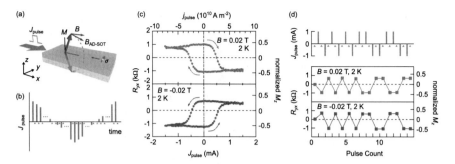

Fig. 4.11 a Schematic illustration of current induced magnetization switching. **b** Measurement procedure. Hall resistance is measured under low current ($\sim 1\,\mu$A) at the green triangles after current pulse injection as shown in the gray bars. The pulse width is ~ 1 ms. **c** Current pulse amplitude J_{pulse} dependence of Hall resistance R_{yx} under in-plane magnetic field $B = \pm 0.02$ T. The current pulse density j_{pulse} is shown on the upper scale. The normalized M_z, calculated from R_{yx}/R_{AHE}, is shown in the right scale, where $R_{\text{AHE}} = 2.3$ kΩ. This corresponds to $(r_{\text{up}} - r_{\text{down}})/(r_{\text{up}} + r_{\text{down}})$, where r_{up} (r_{down}) is the fraction of up (down) domain. **d** Nonvolatile magnetization switching under $B = \pm 0.02$ T and $J_{\text{pulse}} = \pm 1$ mA. Reproduced from [48] with permission from American physical society (Color figure online)

but also to the dc one. Since $R_{\text{AHE}}c_{\text{AD-SOT}} \ll R_{yx}^{(2)}$, the Hall resistance measured under dc current and in-plane magnetization follows $R_{yx} = V_y/J_x \simeq R_{yx}^{(2)}J_x$. Such current-direction-dependent Hall resistance caused by the asymmetric magnon scattering mechanism is difficult to differentiate from AD-SOT [9, 10]. Non-volatile magnetization switching by current pulse injection can avoid this complication by directly accessing the magnetization. Figure 4.11a shows the schematic illustration of magnetization switching; current pulse J_{pulse} is applied parallel to the in-plane magnetic field B. Here, the effective field $B_{\text{AD-SOT}}$ rotates the magnetization with the assistance of small in-plane B. The current pulse injection scheme is shown in Fig. 4.11b. After current pulse injection, the change of M_z is evaluated by Hall resistance R_{yx} measured under low current (1 μA, green triangle) [2]. The measurements were done with a current source (Keithley: Model 6221) and a voltmeter (Keithley: Model 2182A). Figure 4.11c shows $R_{yx} - J_{\text{pulse}}$ loops under in-plane field $B = \pm 0.02$ T. Note that 0.02 T is much smaller than the magnetic anisotropy and therefore magnetization is pointing almost completely up or down. Thus, the intermediate value of normalized M_z originates from the multidomain formation. As can be clearly seen from Fig. 4.11c, R_{yx} is switched from positive (negative) to negative (positive) at around $J_{\text{pulse}} = +0.5$ mA (-0.5 mA) under $B = 0.02$ T, pointing to the current induced magnetization switching. The switching direction is reversed under $B = -0.02$ T. Moreover, the magnetization direction is controlled repeatedly by current pulse of ± 1 mA as shown in Fig. 4.11d. From these experiments, we conclude that the switching is caused by AD-SOT (Fig. 4.11a). We note that the magnetization switching ratio (normalized M_z) is about ~ 0.4, which is smaller than the full switching operation in NM/FM heterostructures [2–5], whose origin is yet to be clarified. At present, we speculate that some part of the domains are hard to switch by current pulse due to the strong pinning. The threshold current $J_{th} = 0.5$ mA corresponds to current density of 2.5×10^{10} A m^{-2}. Despite the larger perpendicular anisotropy, the threshold current density is much smaller than those ($10^{11} - 10^{12}$ A m^{-2}) of NM/FM heterostructures [2–5]. This can be attributed to the smaller saturation magnetization $M_s = 3.5 \times 10^4$ A m^{-1} as well as to the larger θ_{SH} of TI.

In fact, we can estimate the spin Hall angle from the threshold current for magnetization switching. According to the macrospin model, when $K \gg B$, the threshold current density is given by

$$ j_{th} = \frac{2e}{\hbar} \frac{M_s t_{CBST}}{\theta_{SH}} \left(\frac{K}{2} - \frac{B}{\sqrt{2}} \right), \tag{4.16} $$

where t_{CBST} is the thickness of the CBST layer [3]. Using this formula and putting $j_{th} = 2.5 \times 10^{10}$ A/m^2, we can estimate the upper limit for θ_{SH} to be ~ 4. Note that this value is usually overestimated by about one order of magnitude, because actual threshold current gets lower than that estimated by macrospin model due to heating and/or multi-domain formation [3, 56, 57]. However, the estimation of spin Hall angle of order of unity by current-pulse-induced magnetization switching is consistent with other estimations [11–14], showing a high potential of the TI heterostructures as spintronic materials.

We note that a recent paper studies a current-induced magnetization switching in a magnetic insulator in a similar way to our measurement [55]. The result shows a good consistency with our measurement. The required magnitude of current for switching is similar to our measurement for a width of the Hall bar of around 20 μm (10 μm in our case).

4.5   Topological Hall Effect and Skyrmion Formation

The geometry and topology in the Hilbert space is a central issue in quantum physics and recently sheds a new light also on the electronic states in condensed matter. The wavefunctions are characterized by the Berry connection between the two neighboring points in momentum and in real spaces, which works as a vector potential in a given space. The pseudo vector potential mimics electromagnetism leading to the concept of emergent electromagnetic field (EEMF). Topological integers represent the global topology of the manifold in Hilbert space. Since integers cannot continuously change, it gives topological stability to the system. For example, Chern number, which is given by the integral of the Berry curvature over the Brillouin zone, protects the surface/edge states, i.e., bulk-boundary correspondence. The one-dimensional chiral edge modes in quantum Hall effect, and two-dimensional Dirac surface states in three-dimensional topological insulators (TI) [29, 58]are the representative examples of this phenomenon. Especially, as mentioned in the previous Chapter, it has recently been theoretically proposed [59, 60] and observed [49, 50, 61, 62] that TI with doped magnetic ions, i.e., Cr or V, produces the quantized version of anomalous Hall effect (AHE) with no external magnetic field.

Topology in real space, meanwhile, is exemplified by skyrmion spin texture [63–65] found in chiral magnets such as MnSi [63] or $Fe_{1-x}Co_xSi$ [64] etc. Here, the solid angle subtended by the spins gives rise to an emergent magnetic field in real space, and its integral over the two-dimensional space defines the topological integer called skyrmion number. In other words, the skyrmion number counts the number of times the spin direction wraps around a unit sphere. This integer protects the skyrmion from annihilation and allows it to behave as a quasiparticle.

So far, the EEMF in momentum and real spaces have been investigated separately. Recent advances in fabricating artificial structures of materials, however, enable the design of emergent physics containing both real and momentum spaces. Here, we examine these geometrical Hall effects in TI heterostructures composed of magnetic TI $Cr_x(Bi_{1-y}Sb_y)_{2-x}Te_3$ (CBST) and nonmagnetic TI $(Bi_{1-x}Sb_x)_2Te_3$ (BST) [24, 49, 50, 61, 66, 67], which were grown on InP(111) substrates using molecular-beam epitaxy [24, 24, 67]. We control the Fermi energy (E_F) of the TI heterostructures in a field-effect transistor structure (FET) in a broad energy range over the bulk band gap. Especially, the quantum anomalous Hall effect realized in magnetic TIs offers a unique platform where two different physical mechanisms coexist, i.e., the skyrmion formation due to the doping of carrier into the quantum Hall ferromagnet [68] and the skyrmion formation by the interfacial Dzyaloshinskii-Moriya (DM) interaction [69–72].

First, we investigate the gate voltage V_G and temperature T dependence of Hall effect for CBST (2 nm)/BST (5 nm) heterostructure, as shown in Fig. 4.12a. The hysteresis of the Hall conductivity σ_{xy} gradually changes shape with V_G at 2 K (Fig. 4.12b), taking the maximum at $V_G = 0.2$ V. We study the temperature dependence at $V_G = 0.2$ V and -7.0 V in Fig. 4.12c, d. Different from the conventional temperature dependence of magnetic TI at $V_G = 0.2$ V (Fig. 4.12c) where E_F is near the Dirac point, we find two notable features under hole accumulation at $V_G = -7.0$ V in Fig. 4.12d: (i) sign reversal of the anomalous Hall effect σ_{xy}^A under 0 T at 6 K and (ii) non-monotonous behavior in the σ_{xy} to the magnetic field. We here demonstrate that we can understand these phenomena both from the viewpoint of geometrical Hall effects. Detailed data of σ_{xy}^A as a function of T and V_G (Fig. 4.12e) shows the sign reversal at around negative V_G. Such a sign reversal of σ_{xy}^A does not appear in single layer CBST, which always gives the positive sign regardless of its E_F position and temperature [49, 50, 61], as exemplified in Fig. 4.14a. Hence, such a transition between positive and negative σ_{xy}^A is a distinct behavior in the heterostructures of 2-nm/5-nm (Fig. 4.12d) and 3-nm/5-nm (Fig. 4.14b). Note that the longitudinal conductivity of about 10^2 $(\Omega \, cm)^{-1}$ is so low that each anomalous components are not likely to be extrinsic origin such as skew scattering; the intrinsic origin should be explored in the electronic band structure characteristic of the TI heterostructures.

To discuss the sign reversal of anomalous Hall conductance, we theoretically study the low-energy effective Hamiltonian of the TI heterostructures. We consider two examples, 10-nm CBST single layer and 5-nm CBST/BST (5 nm) heterostructure,

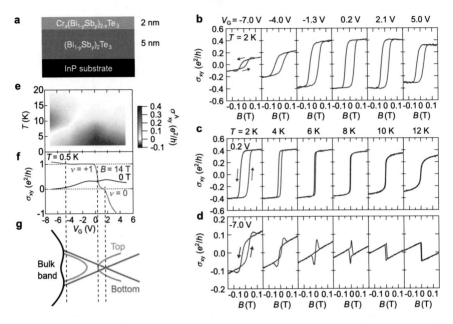

Fig. 4.12 Anomalous Hall conductivity of a topological insulator (TI) heterostructure. **a** A schematic illustrwation of a TI heterostructure. **b** Magnetic field dependence of the Hall conductivity σ_{xy} under various gate voltages V_G at $T = 2$ K. Red (blue) curve represents magnetic field forward (backwar) direction. **c** and **d** Magnetic field dependence of σ_{xy} under various temperatures at $V_G = 0.2$ V (**c**) and $V_G = -7.0$ V (**d**), respectively. **e** A 2D plot of anomalous Hall conductivity σ_{xy}^A as a function of V_G and temperature. Red (Blue) region corresponds to positive (negative) σ_{xy}^A, respectively. **f,** Hall conductivity as a function of V_G at $T = 0.5$ K with $B = 0$ T (black) and with $B = 14$ T (red), showing $\nu = +1$ and $\nu = 0$ quantum Hall plateaus. **g,** A schematic band structure of the heterostructure. Reprinted/Adapted/Translated by permission from Springer Nature: Nature [25], COPYRIGHT (2016) (Color figure online)

as represented in Fig. 4.13a, d. The obtained band structure of CBST (Fig. 4.13b) shows gapped Dirac surface bands in addition to the bulk bands. In the case of CBST/BST (Fig. 4.13e), the Dirac band at the top surface of CBST is gapped due to the exchange coupling while the one at the bottom surface of BST remains gapless. Moreover, the top of the valence bands shows Rashba splitting due to the broken spatial inversion symmetry in the heterostructures (red lines in the lower inset of Fig. 4.13e), and interestingly, these Rashba split bands are slightly gapped due to the exchange coupling (black lines). Such Rashba splitting in the bulk band of TI is also identified in previous ARPES studies [73]. In Fig. 4.13c, f, we calculate the Hall conductivity obtained from the Berry curvature that reflects the nontrivial geometry in the momentum space. In both cases, σ_{xy} is maximized when E_F is within the gap of the surface bands. Noticeably, σ_{xy} in Fig. 4.13f shows a sharp negative peak at the top of the valence bands as denoted by a red triangle. This is attributed to the slightly gapped Rashba split bands on which the Berry curvature is concentrated. Moreover, the negative peak decreases with larger exchange coupling, namely, as

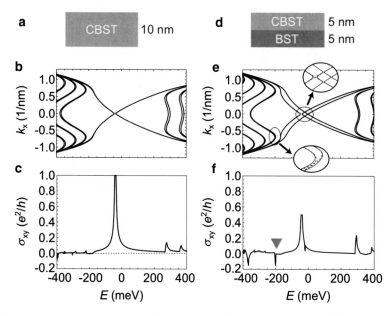

Fig. 4.13 Calculation of Hall conductivity. **a** and **d** Schematic illustration of CBST (10 nm) (**a**) and CBST(5 nm)/BST(5 nm) heterostructure (**d**). **b** and **e** The calculated band structures for a single layer CBST (**b**) and a CBST/BST heterostructure (**e**), where k_x represents the in-plane momentum. **c** and **f** The calculated anomalous Hall conductivity as a function of the Fermi energy E for the single-layer CBST (**c**) and the CBST/BST heterostructure (**f**). The red triangle emphasizes the position of a negative contribution to the anomalous Hall conductivity. Reprinted/Adapted/Translated by permission from Springer Nature: Nature [25], COPYRIGHT (2016) (color figure online)

the temperature gets lowered, the negative contribution of the anomalous Hall effect increases. These theoretical arguments coincide well with the experimental results, as shown in Fig. 4.12e, g. The band energy scheme is based on the results of σ_{xy} at zero and high (14 T) fields including $\nu = 1$ and $\nu = 0$ quantum Hall plateaus (Fig. 4.12f) [24].

In addition to the sign reversal, we observe an anomaly in magnetic field dependence in the Hall effect in CBST (2 nm)/BST (5 nm) heterostructures, as shown in Fig. 4.14a. To clarify the nature of the hysteretic anomaly in the magnetic field dependence, we compare R'_{yx} ($= R_{yx} - R_0 B$) with the magnetization of the film in the top panel of Fig. 4.14d. Here, R_0 is the ordinary Hall coefficient determined from the linear slope at a high magnetic field. In the conventional theory, R'_{yx} is corresponds to the anomalous Hall term, which is proportional to the magnetization M. In our experiment, however, a large discrepancy between R'_{yx} and M is discerned; we need to consider the nontrivial Hall component (shown in Fig. 4.14b bottom) in addition to the M-linear anomalous Hall term to reproduce the observed R'_{yx}. As a plausible origin, we propose skyrmion formation from the theoretical and experimental points of view. Under skyrmion spin texture, the moving electrons feel EEMF, giving rise to the additional Hall component termed topological Hall effect (THE, note that "topo-

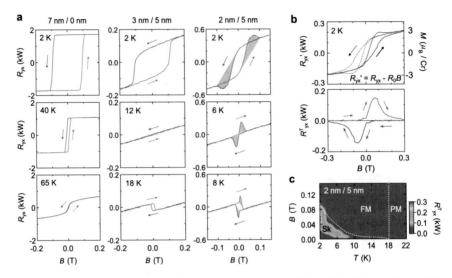

Fig. 4.14 Topological Hall effects of TI heterostructures. **a–c**, Magnetic field dependence of the Hall resistance at various temperatures for a single layer CBST(7 nm), CBST (3 nm)/BST (5 nm) heterostructure, and CBST (2 nm)/BST (5 nm) heterostructure. **b** Upper panel shows the magnetic field dependence of R'_{yx} (Hall resistance subtracted by ordinarily Hall resistance, brown line) in comparison with the magnetization M (blue line) for CBST (2 nm)/BST (5 nm) heterostructure. The lower panel represents the topological Hall component R^T_{yx}. The Red (blue) curve represents the magnetic field increasing (decreasing) process. **c** Magnetic field, and temperature dependence of R^T_{yx}. Sk, FM, and PM stand for skyrmion, ferromagnet, and paramagnet, respectively. FM and PM states were determined from the temperature dependence of Hall effect. Reprinted/Adapted/Translated by permission from Springer Nature: Nature [25], COPYRIGHT (2016) (color figure online)

logical" is defined in real space in the present case) [74, 75], which is experimentally observed in the skyrmion phase of some chiral-lattice magnets [76, 77] as well as in the frustrated magnets endowed with scalar spin chirality [78, 79].

To confirm the possibility of skyrmion formation, we simulated the energetic stability of a skyrmion using a three-dimensional tight-binding model. We consider following three cases; Bloch (Fig. 4.15a), Néel1 (Fig. 4.15b) and Néel2 (Fig. 4.15c) for CBST single layer (Fig. 4.15d) and CBST/BST heterostructure (Fig. 4.15h). In Fig. 4.15e–g, i–k, we plot the formation energy of a skyrmion relative to that of the spin-collinear ferromagnet as a function of the skyrmion radius R at various doping levels. The energy is always minimized at $R = 0$ in the CBST single layer in Fig 4.15e–g, meaning that the ferromagnetic state is the ground state regardless of E_F. For the electron-doped (Fig. 4.15i) or half-filling (Fig. 4.15j) cases of CBST/BST heterostructure, the situation is similar. In the hole-doped case in the heterostructure (Fig. 4.15k), on the other hand, a negative energy region exists for the Néel 2 type skyrmion ($R \neq 0$) denoted by a red triangle. We can explain the appearance of the stable condition for skyrmion formation as follows. The surface state of TI presents inversion symmetry breaking and strong spin-orbit interaction, as exemplified by the

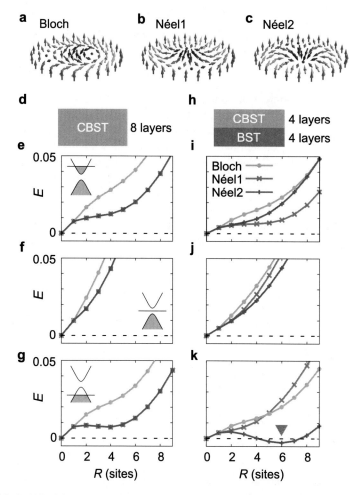

Fig. 4.15 Stability of a skyrmion on magnetic TIs. **a–c** Schematic illustrations of Bloch (**a**), Néel1 (**b**) and Néel2 (**c**) type skyrmions. **d** Schematic illustrations of CBST single layer with 8 layers. **e–g**, The calculated energy of a single skyrmion relative to that of the collinear ferromagnet for electron-doped case (**e**), half-filling case (**f**), and hole-doped case (**g**) for the single-layer structure shown in d as a function of skyrmion radius R. **h** Schematic illustrations of CBST/BST heterostructure with 4 layers each. **i–k** The same as in **e–g** for electron-doped case (**i**), half filling case (**j**) and hole-doped case (**k**) for the heterostructure shown in h, as a function of skyrmion radius R. The red triangle shows the optimized R to stabilize the skyrmion formation. Green, red and blue lines in **e–g** and **i–k** correspond to Bloch, Néel1, and Néel2 type skyrmion, respectively. R is measured in a unit of lattice spacing. Reprinted/Adapted/Translated by permission from Springer Nature: Nature [25], COPYRIGHT (2016) (Color figure online)

spin-momentum-locked Dirac surface state. Therefore, the electrons at the surface states will mediate DM interaction where the DM vector points along the in-plane direction [69–72, 80]. This in-plane DM vector at the surface state of magnetic TI facilitates the formation of Néel type skyrmion. Meanwhile, in the single-layer CBST, DM vectors with opposite signs exist both at the top and the bottom surfaces. This disfavors the skyrmion formation because of the frustration. This is the reason why the heterostructure with broken spatial inversion symmetry is the essential ingredient for the skyrmion formation.

Theoretical verification of skyrmion formation at the hole-accumulated condition coincides well with the experimental situation at large negative V_G (-7.0 V), as shown in Fig. 4.12d. In 2-nm/5-nm heterostructures (Fig. 4.14a), the topological Hall components R_{yx}^T are identified as the green region, which is displayed in Fig. 4.14c against magnetic field and temperature. We can see that skyrmions are formed in the course of the magnetization reversal in a wide temperature range below Curie temperature. The maximum THE is around 140 Ω (the lower panel of Fig. 4.14d), which corresponds to the magnitude of emergent magnetic flux density of 0.29 T using ordinarily Hall coefficient of 490 Ω/T. On the other hand, no THE is observed in a CBST (3 nm)/BST (5 nm) heterostructure, as shown in Fig. 4.14a, although the sign reversal of AHE is apparent. This observation supports the importance of the DM interaction; as the CBST thickness gets larger, the relative strength of the exchange interaction to the DM interaction gets larger so that skyrmion formation would become less favored.

We note that a recent paper pointed out the possibility that the topological Hall effect might be explained by the sum of the two anomalous Hall components with opposite sign [81]. Although we observe skyrmion-like magnetic texture with a magnetic force microscope in a relevant structure of CBST(4 nm)/BST(5 nm), as shown in the next section, such a possibility is to be explored further in the future.

4.6 Current-Induced Skyrmion Motion

The estimation of DM interaction in metallic systems has been a long-standing issue from the 1980s [82]. Recently, the DM interaction in the metallic system is again in the limelight because of the development of theoretical calculation [80, 83, 84] as well as experimental methods to directly [85–87] or indirectly [88] measure the DM interaction. Interestingly, the DM interaction of the metallic system may be manipulated by changing the Fermi level of the system. For example, in $Mn_{1-x}Fe_xGe$, the sign reversal of DM interaction has been observed through the helicity reversal of skyrmion [88]. The theoretical calculation demonstrated the importance of the band crossing point (Weyl point) as a source of DM interaction [84]. However, the band structure in $Mn_{1-x}Fe_xGe$ is too complicated to discuss the contribution of each band crossing point. Here, TI with a single Dirac point gives an ideal playground for studying such effect. In fact, it is theoretically predicted that the DM interaction changes sign as Fermi level goes across the Dirac point [80]. Thus, the experimental study of the DM interaction on the surface of TI would give deep insight into the DM interaction in itinerant systems.

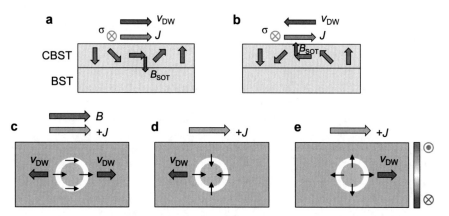

Fig. 4.16 a Schematic illustration of current-induced motion of Néel type domain wall seen from the side of the thin film. Current injection at the top surface of CBST/BST induces spin accumulation σ and causes the effective field by spin-orbit torque $B_{SOT} \propto \sigma \times m$. The spin-orbit torque induces the domain wall motion along the current direction. **a** Schematic illustration of current-induced motion of Néel type domain wall with opposite helicity. In this case, spin-orbit torque works in the opposite direction so that domain wall moves along the opposite to the current direction. **c** Schematic illustration of the current induced motion of domain wall seen from the top of the thin film. The red (blue) region represents the magnetization direction up (down). The down magnetic domain expands under in-plane parallel magnetic field and current. **d, e** Current-induced motion of Néel type skyrmion under zero magnetic field. Depending on the helicity of the skyrmion, skyrmion move parallel or anti-parallel to the current direction (Color figure online)

In order to investigate the DM interaction in magnetic TI, we utilize the current-induced motion of magnetic skyrmion. Recently, current-induced skyrmion motion gains keen interest in spintronics as an efficient method to electrically manipulate these topologically protected structures for a practical application as a nonvolatile memory. In fact, such current-induced motion has already been observed in NM/FM heterostructures such as Ta/CoFeB [71, 89–91]. Here, by combining the skyrmion formation and current-induced magnetization switching, as discussed in the previous sections, we can achieve the current-induced skyrmion motion. In the below, we briefly explain the mechanism for the current-induced motion of Néel type domain wall and Néel type skyrmion by spin-orbit torque and why we can estimate the sign of DM interaction from the current-induced motion of skyrmion. As shown in Fig. 4.16a, current-injection induces the spin-accumulation σ at the surface state, which causes the effective field by spin-orbit torque $B_{SOT} \propto \sigma \times m$ as explained in the previous sections. As a result, magnetization is tilted to the down direction, and it results in the current-induced motion of Néel type domain wall. Here, if the helicity of the Néel type domain wall is opposite, the skyrmion motion direction becomes opposite because B_{SOT} works in the opposite way (Fig, 4.16b). Figure 4.16a–c show the current-induced motion of skyrmion and domain wall seen from the top of the thin film. When in-plane magnetic field is applied, the magnetization direction at the domain wall points in the magnetic field direction due to the Zeeman field. As a result, current-injection induces the domain wall motion so that down domain expands, as shown in Fig. 4.16c. Meanwhile, when the external magnetic field is zero, $B = 0$ T, the internal magnetic structure at the domain wall is determined by the sign of the DM interaction, as shown in Fig. 4.16d. In this case, the current injection does not induce the expansion of the domain. Instead, it induces the translational motion of skyrmion opposite to the current direction. When the helicity of the skyrmion is opposite, the skyrmion motion direction is reversed, as shown in Fig, 4.16e. Thus, we can determine the helicity of Néel type skyrmion and the sign of the DM interaction from the direction of the current-induced motion of skyrmion and the sign of the spin Hall angle [6, 7].

First, we investigate the Fermi level dependence of the sign of the spin Hall angle in $Cr_x(Bi_{1-y}Sb_y)_{2-x}Te_3$ (4 nm)/$(Bi_{1-x}Sb_x)_2Te_3$ (5 nm) samples. Note that we made the CBST layer thicker than that in the previous section since a large magnetic moment is required for the observation by MFM. By changing the composition y, we can go above or below the Dirac point, as shown in Fig. 4.17a, b. We can confirm that the ordinary Hall effect is negative for sample A, while it is positive for sample B, namely the Fermi level is electron (hole) side for sample A (B). By injecting the current parallel to the magnetic field, Hall resistance changes, pointing to the current-induced magnetization switching. For both sample A and sample B, the down domain expands as we apply the in-plane parallel magnetic field and current. This means that the sign of the spin Hall angle of both samples is the same. This is consistent with the previous study of spin-torque ferromagnetic resonance (ST-FMR) on topological insulator [14].

Having studied the current-induced magnetization switching, we next see the current-induced motion of skyrmion in sample A. As mentioned in the previous section, skyrmions are formed at the magnetization reversal in CBST/BST due to the DM interaction at the surface state. In Fig. 4.18a, we can see the skyrmion bubble with diameters of about 500 nm. The injection of the current pulse causes the skyrmion motion opposite to the current direction. This can also be confirmed by taking the correlation function before and after the current injection, as shown in Fig. 4.18c, d. If the skyrmion motion did not occur, the peak of the correlation function would be at the center $(\Delta x, \Delta y) = (0 \text{ nm}, 0 \text{ nm})$. Meanwhile, we can see that the peak position is shifted to around $(\Delta x, \Delta y) = (-130 \text{ nm}, 0 \text{ nm})$. This confirms the current-induced motion of skyrmion in CBST/BST. The skyrmion motion direction becomes opposite for the opposite current direction, as shown in Fig. 4.18b, d. Here, the skyrmion Hall effect, or transverse motion of skyrmion, is not observed. This is probably because the skyrmion motion undergoes the creep motion under the present current density [91]. Figure 4.18e demonstrates the current density dependence of the skyrmion motion speed. We can see that the threshold current density for the skyrmion motion is at around 1.3×10^{11} A/m^2. The skyrmion motion speed increases with the current amplitude after the threshold current. We cannot estimate the skyrmion motion speed above 2.0×10^{11} A/m^2 because large current injection disturbs the magnetic domain structure probably due to heating.

Figure 4.19 compares the current-induced magnetization switching, and current-induced motion of skyrmion in sample A. Since current-induced magnetization

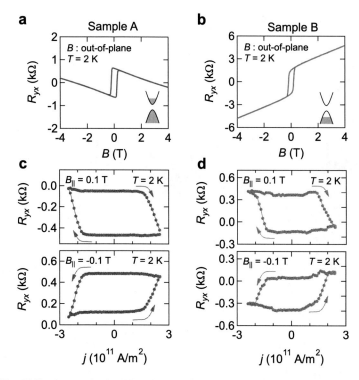

Fig. 4.17 **a** Hall resistance for $Cr_x(Bi_{1-y}Sb_y)_{2-x}Te_3/(Bi_{1-y}Sb_y)_2Te_3$ sample with $y = 0.80$ (electron side, sample A). **b** Hall resistance for $Cr_x(Bi_{1-y}Sb_y)_{2-x}Te_3/(Bi_{1-y}Sb_y)_2Te_3$ sample with $y = 0.92$ (hole side, sample B). **c** Current pulse density j dependence of Hall resistance R_{yx} under in plane magnetic field $B = \pm 0.1$ T at $T = 2$ K for sample A. **d** The same as (**c**) for sample B

switching and current-induced skyrmion motion occurs at the similar critical current density, we can confirm that the spin-orbit torque induces skyrmion motion due to the spin accumulation at the surface state.

The current pulse number dependence of the skyrmion motion distance for sample A is shown in Fig. 4.20a. The motion distance of skyrmion per pulse gets smaller as we increase the current pulse number. This indicates that the skyrmions are trapped to the impurity as the current is applied; when skyrmions are trapped to the local energy minimum at the impurity, they can be no longer moved by the current-induced torque. Here, we compare the result of the current-induced skyrmion motion for sample A and sample B. As can be seen from Fig. 4.20a, b, the skyrmion motion direction is the same for both samples. Considering the fact that the sign of the spin Hall angle for both samples is the same, this result indicates that the helicity of Néel type skyrmion is the same for the electron and the hole side. This is inconsistent with the expectation from theory [80], which predicts the sign reversal of the DM interaction at the electron/hole side. We pose two possibilities as the reasons for the inconsistency between the experiment and the theory. The first possibility is that

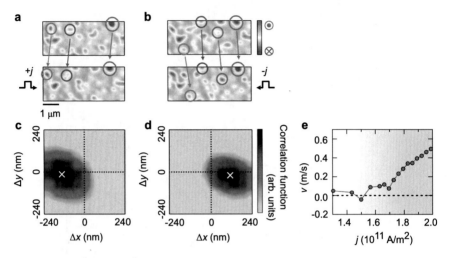

Fig. 4.18 **a** Current-induced motion of skyrmion visualized by MFM. (top) The MFM image before the current injection. (bottom) The MFM image after the successive 50 current pulse injection with a current density of $j = 1.8 \times 10^{11}$ A/m^2 and pulse width of 100 ns. All the MFM measurement is done at $T = 10$ K. **b** The same as a for the opposite current direction. **c** The correlation function before and after the current pulse injection. The peak of the correlation function shifts in a negative direction, meaning that skyrmions moved opposite to the current direction, d, The same as b for the opposite current direction. **e** The current density dependence of the skyrmion motion speed estimated from the single current pulse of 100 ns. The skyrmion motion speed is estimated from the correlation function after the injection of a single pulse with pulse width about 100 ns

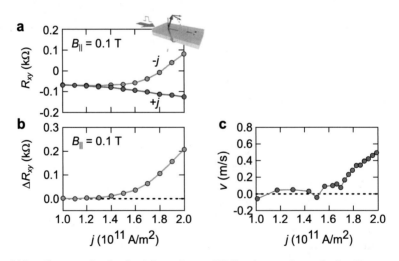

Fig. 4.19 **a** Current pulse density j dependence of Hall resistance R_{yx} under in-plane magnetic field $B_{\parallel} = 0.1$ T for plus current $+j$ (blue) and minus current $-j$ (red). **b** The difference in the Hall resistance ΔR_{yx} for plus and minus current. **c** The current density dependence of the skyrmion motion speed (Color figure online)

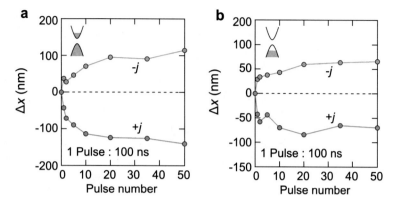

Fig. 4.20 a, b The current pulse number dependence of the skyrmion motion distance Δx for sample A (**a**) and sample B (**b**) under plus (red) and minus (blue) current. The skyrmion motion distance is estimated from the correlation function (Color figure online)

the skyrmion motion is caused not by the spin-orbit torque but by the spin-transfer torque. In this case, the skyrmion motion direction is independent of the helicity of Néel type skyrmion and always moves opposite to the current direction. This is unlikely because current-induced magnetization switching and current-induced skyrmion motion occur at a similar critical current density, as discussed in Fig. 4.19, but this may explain the inconsistency if the spin-orbit torque and spin-transfer torque simultaneously works in this system. Secondly, the original theory in Ref. [80] assumed the electron-hole symmetry at the Dirac point. If the electron-hole symmetry exists, the sign of the DM interaction should be reversed at the Dirac point. However, the actual surface state does not possess electron-hole symmetry because of the k^2 term and hexagonal warping. In addition, the band crossing point exists not only at the surface state but also in the bulk valence band, which may give rise to the additional contribution to the DM interaction [25]. Under this condition, the DM interaction may not reverse sign at the Dirac point. Thus, further theoretical and experimental investigations are required to reveal the Fermi level dependence of the DM interaction fully.

4.7 Summary

In this Chapter, we have investigated the spintronics functionalities of magnetic TI $Cr_x(Bi_{1-y}Sb_y)_{2-x}Te_3$. First, we studied unidirectional (UMR) in magnetic TI. UMR is shown to be several orders of magnitude larger than in other reported systems [18–20]. Moreover, from the detailed temperatur and magnetic field dependence, the origin of UMR is identified to be asymmetric magnon scattering. The improvement of theoretical calculation and the detailed comparison with the experiment remains as future issues [9, 12, 14]. The similar nonlinear transport is observed in a transverse resistance when magnetic field is parallel to the current, namely current-nonlinear Hall effect. It was reveraled that the current-nonlinear Hall effect mainly originates from asymmetric magnon scattering. Thus, asymmetric magnon scattering leads to the overestimation of spin Hall angle by current-nonlinear Hall effect. Next, current-induced magnetization switching is realized in a magnetic TI. The required current for switching is shown to be about one order smaller than other systems. Especially, the current-induced control of magnetization in a magnetic TI will enable current-induced switching of the Chern number and the manipulation of chiral edge states at the domain wall in the future. Finally, we demonstrated the skyrmion formation from the surface state of magnetic TI. The spin-momentum locking results in the appearance of finite DM interaction at the surface state, which leads to the formation of Néel type skyrmion. Also, the current-induced motion of skyrmion was demonstrated with MFM.

In summary, we have revealed the versatile spintronic functionalities in magnetic TI. Originating from the unique spin texture at the Fermi surface, spin-momentum locking, magnetic TI results in a larger spintronic responces compared with NM/FM heterostructures. The room temperature realization of these functionalities in a magnetic TI will open up a route towards the application of these functionalities.

References

1. Slonczewski JC et al (1996) Current-driven excitation of magnetic multilayers. J Magn Magn Mater 159(1):L1
2. Ralph DC, Stiles MD (2008) Spin transfer torques. J Magn Magn Mater 320(7):1190–1216
3. Liu L, Lee OJ, Gudmundsen TJ, Ralph DC, Buhrman RA (2012) Current-induced switching of perpendicularly magnetized magnetic layers using spin torque from the spin hall effect. Phys Rev Lett 109(9):096602
4. Liu L, Pai C-F, Li Y, Tseng HW, Ralph DC, Buhrman RA (2012) Spin-torque switching with the giant spin hall effect of tantalum. Science 336(6081):555–558
5. Pai C-F, Liu L, Li Y, Tseng HW, Ralph DC, Buhrman RA (2012) Spin transfer torque devices utilizing the giant spin hall effect of tungsten. Appl Phys Lett 101(12):122404
6. Emori S, Bauer U, Ahn S-M, Martinez E, Beach GSD (2013) Current-driven dynamics of chiral ferromagnetic domain walls. Nat Mater 12(7):611–616
7. Kwang-Su R, Luc T, See-Hun Y, Stuart P (2013) Chiral spin torque at magnetic domain walls. Nat Nanotechnol 8(7):527
8. Edelstein VM (1990) Spin polarization of conduction electrons induced by electric current in two-dimensional asymmetric electron systems. Solid State Commun 73(3):233–235, 1990
9. Fan Y, Upadhyaya P, Kou X, Lang M, Takei S, Wang Z, Tang J, He L, Chang L-T, Montazeri M et al (2014) Magnetization switching through giant spin-orbit torque in a magnetically doped topological insulator heterostructure. Nature materials 13(7):699–704
10. Yabin F, Xufeng K, Pramey U, Qiming S, Lei P, Murong L, Xiaoyu C, Jianshi T, Mohammad M, Koichi M et al (2016) Electric-field control of spin-orbit torque in a magnetically doped topological insulator. Nat Nanotechnol 11(4):352
11. Jamali M, Lee JS, Jeong JS, Mahfouzi F, Lv Y, Zhao Z, Nikolić BK, Mkhoyan KA, Samarth N, Wang J-P (2015) Giant spin pumping and inverse spin hall effect in the presence of surface and bulk spin- orbit coupling of topological insulator bi2se3. Nano Lett 15(10):7126–7132

12. Mellnik AR, Lee JS, Richardella A, Grab JL, Mintun PJ, Fischer MH, Vaezi A, Manchon A, Kim E-A, Samarth N et al (2014) Spin-transfer torque generated by a topological insulator. Nature 511(7510):449–451

13. Yi W, Praveen D, Karan B, Nikesh K, Matthew B, Oh S, Hyunsoo Y (2015) Topological surface states originated spin-orbit torques in bi 2 se 3. Phys Rev Lett 114(25):257202

14. Kondou K, Yoshimi R, Tsukazaki A, Fukuma Y, Matsuno J, Takahashi KS, Kawasaki M, Tokura Y, Otani Y (2016) Fermi-level-dependent charge-to-spin current conversion by dirac surface states of topological insulators. Nat Phys 12(11):1027–1031

15. Chumak AV, Vasyuchka VI, Serga AA, Hillebrands B (2015) Magnon spintronics. Nat Phys 11(6):453–461

16. Liu L, Moriyama T, Ralph DC, Buhrman RA (2011) Spin-torque ferromagnetic resonance induced by the spin hall effect. Phys Rev Lett 106(3):036601

17. Miron IM, Garello K, Gaudin G, Zermatten P-J, Costache MV, Auffret S, Bandiera S, Rodmacq B, Schuhl A, Gambardella P (2011) Perpendicular switching of a single ferromagnetic layer induced by in-plane current injection. Nature 476(7359):189–193

18. Olejník K, Novák V, Wunderlich J, Jungwirth T (2015) Electrical detection of magnetization reversal without auxiliary magnets. Phys Rev B 91(18):180402

19. Avci CO, Garello K, Ghosh A, Gabureac M, Alvarado SF, Gambardella P (2015) Unidirectional spin hall magnetoresistance in ferromagnet/normal metal bilayers. Nat Phys 11(7):570–575

20. Avci CO, Garello K, Mendil J, Ghosh A, Blasakis N, Gabureac M, Trassin M, Fiebig M, Gambardella P (2015) Magnetoresistance of heavy and light metal/ferromagnet bilayers. Appl Phys Lett 107(19):192405

21. Baibich MN, Broto JM, Fert A, Van Dau FN, Petroff F, Etienne P, Creuzet G, Friederich A, Chazelas J (1988) Giant magnetoresistance of (001) fe/(001) cr magnetic superlattices. Phys Rev Lett 61(21):2472

22. Binasch G, Grünberg P, Saurenbach F, Zinn W (1989) Enhanced magnetoresistance in layered magnetic structures with antiferromagnetic interlayer exchange. Phys Rev B 39(7):4828

23. Hasan MZ, Kane CL (2010) Colloquium: topological insulators. Rev Mod Phys 82(4):3045

24. Yoshimi R, Yasuda K, Tsukazaki A, Takahashi KS, Nagaosa N, Kawasaki M, Tokura Y (2015) Quantum hall states stabilized in semi-magnetic bilayers of topological insulators. Nat Commun 6(1):1–6

25. Yasuda K, Wakatsuki R, Morimoto T, Yoshimi R, Tsukazaki A, Takahashi KS, Ezawa M, Kawasaki M, Nagaosa N, Tokura Y (2016) Geometric hall effects in topological insulator heterostructures. Nat Phys 12(6):555–559

26. Zhang J, Chang C-Z, Zhang Z, Wen J, Feng X, Li K, Liu M, He K, Wang L, Chen X et al (2011) Band structure engineering in (bi 1–x sb x) 2 te 3 ternary topological insulators. Nat Commun 2(1):1–6

27. Chang C-Z, Zhang J, Liu M, Zhang Z, Feng X, Li K, Wang L-L, Chen X, Dai X, Fang Z et al (2013) Thin films of magnetically doped topological insulator with carrier-independent long-range ferromagnetic order. Adv Mater 25(7):1065–1070

28. Yoshimi R, Tsukazaki A, Kozuka Y, Falson J, Takahashi KS, Checkelsky JG, Nagaosa N, Kawasaki M, Tokura Y (2015) Quantum hall effect on top and bottom surface states of topological insulator (bi 1–x sb x) 2 te 3 films. Nat Commun 6(1):1–6

29. Zhang H, Liu C-X, Qi X-L, Dai X, Fang Z, Zhang S-C (2009) Topological insulators in bi 2 se 3, bi 2 te 3 and sb 2 te 3 with a single dirac cone on the surface. Nat Phys 5(6):438–442

30. Xia Y, Qian D, Hsieh D, Wray L, Pal A, Lin H, Bansil A, Grauer DHYS, Hor YS, Cava RJ et al (2009) Observation of a large-gap topological-insulator class with a single dirac cone on the surface. Nat Phys 5(6):398–402

31. Kenji Yasuda, Atsushi Tsukazaki, Ryutaro Yoshimi, KS Takahashi, Masashi Kawasaki, and Yoshinori Tokura. Large unidirectional magnetoresistance in a magnetic topological insulator. *Physical review letters*, 117(12):127202, 2016

32. Zhang Y, He K, Chang C-Z, Song C-L, Wang L-L, Chen X, Jia J-F, Fang Z, Dai X, Shan W-Y et al (2010) Crossover of the three-dimensional topological insulator bi 2 se 3 to the two-dimensional limit. Nat Phys 6(8):584–588

33. Xu S-Y, Neupane M, Liu C, Zhang D, Richardella A, Wray LA, Alidoust N, Leandersson M, Balasubramanian T, Sánchez-Barriga J et al (2012) Hedgehog spin texture and berry's phase tuning in a magnetic topological insulator. Nat Phys 8(8):616–622
34. Ken-Ichi U, Hiroto A, Takeru O, Hiroyasu N, Sadamichi M, Eiji S (2010) Observation of longitudinal spin-seebeck effect in magnetic insulators. Appl Phys Lett 97(17):172505
35. Kikkawa T, Uchida K, Shiomi Y, Qiu Z, Hou D, Tian D, Nakayama H, Jin X-F, Saitoh E (2013) Longitudinal spin seebeck effect free from the proximity nernst effect. Phys Rev Lett 110(6):067207
36. Takashi K, Ken-Ichi U, Shunsuke D, Zhiyong Q, Yuki S, Eiji S (2015) Critical suppression of spin seebeck effect by magnetic fields. Phys Rev B 92(6):064413
37. Onose Y, Ideue T, Katsura H, Shiomi Y, Nagaosa N, Tokura Y (2010) Observation of the magnon hall effect. Science 329(5989):297–299
38. Flipse J, Dejene FK, Wagenaar D, Bauer GEW, Youssef JB, Van Wees BJ (2014) Observation of the spin peltier effect for magnetic insulators. Phys Rev Lett 113(2):027601
39. Saitoh E, Ueda M, Miyajima H, Tatara G (2006) Conversion of spin current into charge current at room temperature: Inverse spin-hall effect. Appl Phys Lett 88(18):182509
40. Shiomi Y, Nomura K, Kajiwara Y, Eto K, Novak M, Segawa K, Ando Y, Saitoh E (2014) Spin-electricity conversion induced by spin injection into topological insulators. Phys Rev Lett 113(19):196601
41. Ashcroft NW, Mermin ND et al (1976) Solid State Phys
42. Skrotskii GV, Kurbatov LV (1966) Ferromagn Reson
43. Avci CO, Mendil J, Beach GSD, Gambardella P (2018) Origins of the unidirectional spin hall magnetoresistance in metallic bilayers. Phys Rev Lett 121(8):087207
44. Kab-Jin Kim, Tian Li, Sanghoon Kim, Takahiro Moriyama, Tomohiro Koyama, Daichi Chiba, Kyung-Jin Lee, Hyun-Woo Lee, Teruo Ono (2019) Possible contribution of high-energy magnons to unidirectional magnetoresistance in metallic bilayers. Applied Physics Express 12(6):063001
45. Yin Y, Han D-S, de Jong MCH, Lavrijsen R, Duine RA, Swagten HJM, Koopmans B (2017) Thickness dependence of unidirectional spin-hall magnetoresistance in metallic bilayers. Appl Phys Lett 111(23):232405
46. Fan Y, Shao Q, Pan L, Che X, He Q, Yin G, Zheng C, Yu G, Nie T, Masir MR et al (2019) Unidirectional magneto-resistance in modulation-doped magnetic topological insulators. Nano Lett 19(2):692–698
47. Langenfeld S, Tshitoyan V, Fang Z, Wells A, Moore TA, Ferguson AJ (2016) Exchange magnon induced resistance asymmetry in permalloy spin-hall oscillators. Appl Phys Lett 108(19):192402
48. Yasuda K, Tsukazaki A, Yoshimi R, Kondou K, Takahashi KS, Otani Y, Kawasaki M, Tokura Y (2017) Current-nonlinear hall effect and spin-orbit torque magnetization switching in a magnetic topological insulator. Phys Rev Lett 119(13):137204
49. Xufeng K, Shih-Ting G, Yabin F, Lei P, Murong L, Ying J, Qiming S, Tianxiao N, Koichi M, Jianshi T et al (2014) Scale-invariant quantum anomalous hall effect in magnetic topological insulators beyond the two-dimensional limit. Phys Rev Lett 113(13):137201
50. Chang C-Z, Zhang J, Feng X, Shen J, Zhang Z, Guo M, Li K, Yunbo O, Wei P, Wang L-L et al (2013) Experimental observation of the quantum anomalous hall effect in a magnetic topological insulator. Science 340(6129):167–170
51. Kim J, Sinha J, Hayashi M, Yamanouchi M, Fukami S, Suzuki T, Mitani S, Ohno H (2013) Layer thickness dependence of the current-induced effective field vector in ta| cofeb| mgo. Nat Mater 12(3):240–245
52. Garello K, Miron IM, Avci CO, Freimuth F, Mokrousov Y, Blügel S, Auffret S, Boulle O, Gaudin G, Gambardella P (2013) Symmetry and magnitude of spin-orbit torques in ferromagnetic heterostructures. Nat Nanotechnol 8(8):587
53. Masamitsu H, Junyeon K, Michihiko Y, Hideo O (2014) Quantitative characterization of the spin-orbit torque using harmonic hall voltage measurements. Phys Rev B 89(14):144425

54. Holstein T, Hl Primakoff (1940) Field dependence of the intrinsic domain magnetization of a ferromagnet. Phys Rev 58(12):1098
55. Xiaoyu C, Quanjun P, Božo V, Jingyi Z, Lei P, Peng Z, Gen Y, Wu H, Qiming S, Peng D et al (2020) Strongly surface state carrier-dependent spin-orbit torque in magnetic topological insulators. Adv Mater 32(16):1907661
56. Zhang C, Yamanouchi M, Sato H, Fukami S, Ikeda S, Matsukura F, Ohno H (2014) Magnetization reversal induced by in-plane current in ta/cofeb/mgo structures with perpendicular magnetic easy axis. J Appl Phys 115(17):17C714
57. Lee OJ, Liu LQ, Pai CF, Li Y, Tseng HW, Gowtham PG, Park JP, Ralph DC, Buhrman RA (2014) Central role of domain wall depinning for perpendicular magnetization switching driven by spin torque from the spin hall effect. Phys Rev B 89(2):024418
58. Fu L, Kane CL, Mele EJ (2007) Topological insulators in three dimensions. Phys Rev Lett 98(10):106803
59. Rui Y, Zhang W, Zhang H-J, Zhang S-C, Dai X, Fang Z (2010) Quantized anomalous hall effect in magnetic topological insulators. Science 329(5987):61–64
60. Kentaro N, Naoto N (2011) Surface-quantized anomalous hall current and the magnetoelectric effect in magnetically disordered topological insulators. Phys Rev Lett 106(16):166802
61. Checkelsky JG, Yoshimi R, Tsukazaki A, Takahashi KS, Kozuka Y, Falson J, Kawasaki M, Tokura Y (2014) Trajectory of the anomalous hall effect towards the quantized state in a ferromagnetic topological insulator. Nat Phys 10(10):731–736
62. Chang C-Z, Zhao W, Kim DY, Zhang H, Assaf BA, Heiman D, Zhang S-C, Liu C, Chan MHW, Moodera JS (2015) High-precision realization of robust quantum anomalous hall state in a hard ferromagnetic topological insulator. Nat Mater 14(5):473–477
63. Mühlbauer S, Binz B, Jonietz F, Pfleiderer C, Rosch A, Neubauer A, Georgii R, Böni P (2009) Skyrmion lattice in a chiral magnet. Science 323(5916):915–919
64. Yu XZ, Onose Y, Kanazawa N, Park JH, Han JH, Matsui Y, Nagaosa N, Tokura Y (2010) Real-space observation of a two-dimensional skyrmion crystal. Nature 465(7300):901–904
65. Naoto N, Yoshinori T (2013) Topological properties and dynamics of magnetic skyrmions. Nat Nanotechnol 8(12):899
66. Kou X, He L, Lang M, Fan Y, Wong K, Jiang Y, Nie T, Jiang W, Upadhyaya P, Xing Z et al (2013) Manipulating surface-related ferromagnetism in modulation-doped topological insulators. Nano Lett 13(10):4587–4593
67. Yoshimi R, Tsukazaki A, Kikutake K, Checkelsky JG, Takahashi KS, Kawasaki M, Tokura Y (2014) Dirac electron states formed at the heterointerface between a topological insulator and a conventional semiconductor. Nat Mater 13(3):253–257
68. Girvin SM (1999) The quantum hall effect: novel excitations and broken symmetries. In: Aspects topologiques de la physique en basse dimension. Topological aspects of low dimensional systems. Springer, pp 53–175
69. Fert A, Cros V, Sampaio J (2013) Skyrmions on the track. Nat Nanotechnol 8(3):152–156
70. Stefan H, Von Bergmann K, Matthias M, Jens B, André K, Roland W, Gustav B, Stefan B (2011) Spontaneous atomic-scale magnetic skyrmion lattice in two dimensions. Nat Phys 7(9):713–718
71. Wanjun J, Pramey U, Wei Z, Yu G, Jungfleisch MB, Fradin FY, Pearson JE, Tserkovnyak Y, Wang KL, Heinonen O et al (2015) Blowing magnetic skyrmion bubbles. Science 349(6245):283–286
72. Moreau-Luchaire C, Moutafis C, Reyren N, Sampaio J, Vaz CAF, Van Horne N, Bouzehouane K, Garcia K, Deranlot C, Warnicke P et al (2016) Additive interfacial chiral interaction in multilayers for stabilization of small individual skyrmions at room temperature. Nat Nanotechnol 11(5):444
73. King PDC, Hatch RC, Bianchi M, Ovsyannikov R, Lupulescu C, Landolt G, Slomski B, Dil JH, Guan D, Mi JL et al (2011) Large tunable rashba spin splitting of a two-dimensional electron gas in bi 2 se 3. Phys Rev Lett 107(9):096802
74. Bruno P, Dugaev VK, Taillefumier M (2004) Topological hall effect and berry phase in magnetic nanostructures. Phys Rev Lett 93(9):096806

75. Tatara G, Kohno H, Shibata J, Lemaho Y, Lee K-J (2007) Spin torque and force due to current for general spin textures. J Phys Soc Jpn 76(5):054707–054707
76. Neubauer A, Pfleiderer C, Binz B, Rosch A, Ritz R, Niklowitz PG, Böni P (2009) Topological hall effect in the a phase of mnsi. Phys Rev Lett 102(18):186602
77. Lee M, Kang W, Onose Y, Tokura Y, Ong NP (2009) Unusual hall effect anomaly in mnsi under pressure. Phys Rev Lett 102(18):186601
78. Taguchi Y, Oohara Y, Yoshizawa H, Nagaosa N, Tokura Y (2001) Spin chirality, berry phase, and anomalous hall effect in a frustrated ferromagnet. Science 291(5513):2573–2576
79. Ueda K, Iguchi S, Suzuki T, Ishiwata S, Taguchi Y, Tokura Y (2012) Topological hall effect in pyrochlore lattice with varying density of spin chirality. Phys Rev Lett 108(15):156601
80. Ryohei W, Motohiko E, Naoto N (2015) Domain wall of a ferromagnet on a three-dimensional topological insulator. Sci Rep 5:13638
81. Wang F, Wang X, Zhao Y-F, Xiao D, Zhou L-J, Liu W, Zhang Z, Zhao W, Chan MHW, Samarth N et al (2020) Berry curvature engineering in magnetic topological insulator heterostructures. arXiv:2004.12560
82. Kataoka M, Nakanishi O, Yanase A, Kanamori J (1984) Antisymmetric spin interaction in metals. J Phys Soc Jpn 53(10):3624–3633
83. Takashi K, Naoto N, Ryotaro A (2015) Control of dzyaloshinskii-moriya interaction in mn 1–x fe x ge: a first-principles study. Sci Rep 5:13302
84. Toru K, Takashi K, Ryotaro A, Gen T (2016) Dzyaloshinskii-moriya interaction as a consequence of a doppler shift due to spin-orbit-induced intrinsic spin current. Phys Rev Lett 116(24):247201
85. Di K, Zhang VL, Lim HS, Ng SC, Kuok MH, Yu J, Yoon J, Qiu X, Yang H (2015) Direct observation of the dzyaloshinskii-moriya interaction in a pt/co/ni film. Phys Rev Lett 114(4):047201
86. Yusuke I, Soichiro U, Kazunori U, Yoshinori O (2015) Nonreciprocal magnon propagation in a noncentrosymmetric ferromagnet life 5 o 8. Phys Rev B 92(18):184419
87. Seki S, Okamura Y, Kondou K, Shibata K, Kubota M, Takagi R, Kagawa F, Kawasaki M, Tatara G, Otani Y et al (2016) Magnetochiral nonreciprocity of volume spin wave propagation in chiral-lattice ferromagnets. Phys Rev B 93(23):235131
88. Shibata K, Yu XZ, Hara T, Morikawa D, Kanazawa N, Kimoto K, Ishiwata S, Matsui Y, Tokura Y (2013) Towards control of the size and helicity of skyrmions in helimagnetic alloys by spin-orbit coupling. Nat Nanotechnol 8(10):723–728
89. Woo S, Litzius K, Krüger B, Im M-Y, Caretta L, Richter K, Mann M, Krone A, Reeve RM, Weigand M et al (2016) Observation of room-temperature magnetic skyrmions and their current-driven dynamics in ultrathin metallic ferromagnets. Nat Mater 15(5):501–506
90. Yu G, Upadhyaya P, Shao Q, Wu H, Yin G, Li X, He C, Jiang W, Han X, Amiri PK et al (2017) Room-temperature skyrmion shift device for memory application. Nano Lett 17(1):261–268
91. Jiang W, Zhang X, Yu G, Zhang W, Wang X, Jungfleisch MB, Pearson JE, Cheng X, Heinonen O, Wang KL et al (2017) Direct observation of the skyrmion hall effect. Nat Phys 13(2):162–169

Chapter 5
Transport Property of Topological Insulator/Superconductor Interface

5.1 Introduction

Topological superconductivity is extensively sought for in contemporary condensed matter physics [1–3]. In an ordinary superconductor, electrons with opposite spin directions form a spin-singlet Cooper pair and even-parity s-wave or d-wave superconductor is realized. In a spinless superconductor with no spin degeneracy, on the other hand, an odd-parity p-wave superconductor is induced due to the Pauli principle. These unique superconductors are topologically non-trivial and support Majorana zero modes at the boundary or the vortex. This can work as building blocks for robust quantum computation owing to their non-Abelian statistics [1–3]. For realizing a non-degenerate spinless mode, one needs to fight against Kramers degeneracy: One way to deal with the Kramers degeneracy is to utilize Rashba-type spin-splitting as realized in superconducting-proximity coupled [4–6] or topological insulator (TI) surface states [7–11]. At the surface or interface, relativistic spin-orbit coupling and inversion symmetry breaking leads to the appearance of spin-momentum locked Dirac surface state. Here, the electron spin is fixed to 90 degrees with respect to momentum direction, so that superconducting-proximity effect realizes a topological superconductor (TSC) [7]. Following the theoretical prediction, many experimental studies have been undertaken to detect the Majorana bound state at the edge/boundary of a material [8, 9] or at the vortex core [11].

Recently, nonreciprocal transport measurement is established as a new probe to detect the broken inversion symmetry in electronic systems [12–20]. The electrical resistivity is expected to vary depending on the current direction, reflecting the asymmetric nature of the spin-split electronic band structure. For example, nonreciprocal resistance is experimentally observed in materials with Rashba spin splitting, such as a bulk Rashba semiconductor [12], TI surfaces [16–18], and heterointerfaces [13–15]. Under the in-plane magnetic field perpendicular to the current, the resistance depends on the spin accumulation direction relative to the external magnetic field, either parallel or antiparallel [12, 15–18]. Besides, nonreciprocal transport is observed in non-centrosymmetric superconductors like a chiral nanotube [19] and a

© Springer Nature Singapore Pte Ltd. 2020
K. Yasuda, *Emergent Transport Properties of Magnetic Topological Insulator Heterostructures*, Springer Theses,
https://doi.org/10.1007/978-981-15-7183-1_5

transition metal dichalcogenide [20], which demonstrate that the inversion symmetry breaking of the crystal is reflected in the Cooper pairs of superconductors. Therefore, the nonreciprocal transport measurement can help study the underlying symmetry of Cooper pairs in TSC candidates.

5.2   Nonreciprocal Charge Transport at Topological Insulator/Superconductor Interface

We target the interface superconductivity between Bi_2Te_3 and FeTe [21]. The interface superconductivity is induced when a 3DTI, Bi_2Te_3. [22] is grown on a superconducting parent compound, FeTe, as shown in Fig. 5.1a. Since the topological surface state resides on the interface, the interaction between the surface state and superconductivity makes this system a strong candidate for TSC. It has remained unclear, however, whether the nature of superconductivity reflects the spin-momentum locking of the surface state [21]. We here measure the nonreciprocal transport to study the symmetry of the superconductivity. Figure 5.1b shows the temperature dependence of resistance for the respective samples, FeTe, Bi_2Te_3 and Bi_2Te_3/FeTe thin films. FeTe thin film shows a semi-metallic behavior with a kinked structure at around 50 K associated with the antiferromagnetic transition [23]. Meanwhile, Bi_2Te_3 exhibits a metallic behavior. The resistance of Bi_2Te_3/FeTe behaves as parallel conduction of these two materials except the resistance drop to zero at around 10 K. This indicates the appearance of superconductivity at the interface [21] (Fig. 5.1c).

We characterize the nature of superconductivity from the resistance measurement. Figure 5.2a shows the temperature dependence of resistance in Bi_2Te_3/FeTe. The gradual decrease of resistance is observed at around the superconducting onset temperature. This feature is well reproduced with the Aslamazov-Larkin pair contribution to conductivity. The appearance of fluctuating Cooper pairs leads to the initiation of a new conducting channel for charge transport, leading to the decrease of resistance as follows [25]:

$$R(T) = \left(\frac{1}{R_N(T)} + \Delta G_{AL} \right)^{-1}, \tag{5.1}$$

where $R_N(T)$ is the temperature dependence of normal resistance and ΔG_{AL} is the excess conductance due to the emerging superconducting channel. We fit the resistance curve by

$$R_N(T) = a + cT, \Delta G_{AL} = \frac{e^2}{16\hbar} \left(\frac{T_{c0}}{T_{c0} - T} \right), \tag{5.2}$$

where T_{c0} the temperature at which the finite amplitude of the order parameter develops. The blue dotted curve in Fig. 5.2a is the fitting by the Aslamazov-Larkin formula,

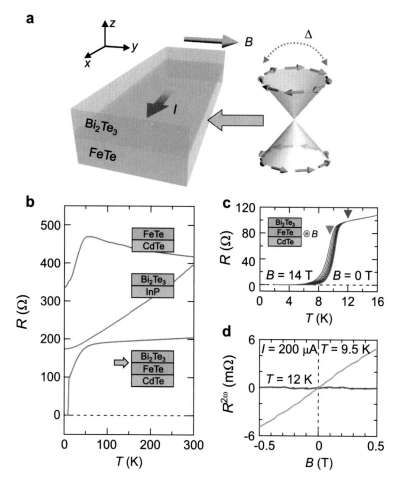

Fig. 5.1 Basic characterization of a Bi_2Te_3/FeTe heterostructure. **a** The illustration of a Bi_2Te_3/FeTe heterostructure (Bi_2Te_3: green, FeTe: blue). The interface of Bi_2Te_3 and FeTe shows superconductivity (yellow region). The nonreciprocal transport is measured with current along y-direction under in-plane B field perpendicular to the current. The interaction between superconductivity (as represented by Δ) and topological surface state with spin-momentum locking is expected at the interface. **b** Temperature dependence of resistance in FeTe(20 nm)/CdTe(100) (green), Bi_2Te_3 (10 nm)/InP(111) (blue) and Bi_2Te_3 (10 nm)/FeTe(20 nm)/CdTe(100) (red) thin films. The superconductivity and topological surface state appear at the interface between Bi_2Te_3 and FeTe as denoted in the yellow arrow. **c** Temperature dependence of resistance in Bi_2Te_3/FeTe heterostructure under in-plane magnetic field $B = 0, 2, 4, 6, 8, 10, 12$ and 14 T. The resistance is measured under $I = 1 \, \mu A$. **d** The magnetic field dependence of 2nd harmonic resistance in Bi_2Te_3/FeTe heterostructure for $T = 12$ K (blue, normal state) and for $T = 9.5$ K (light blue) measured under $I = 20 \, \mu A$. Reprinted from [24], Copyright Author(s) 2019 Limited. Licensed under CC BY 4.0 (Color figure in online)

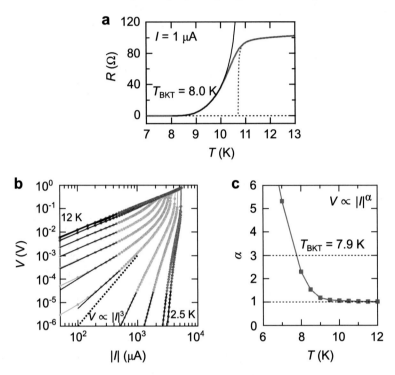

Fig. 5.2 **a** The temperature dependence of resistance measured with the injection of $I = 1\,\mu A$. The dotted blue curve is the fitting result by the Aslamazov-Larkin equation. The solid black curve is the fitting line by the Halperin-Nelson equation. **b** A log-log plot of current-voltage (I-V) characteristics. The measurement temperatures are $T = 2.5, 4, 6, 7, 8, 8.5, 9, 9.5, 10, 10.5, 11, 11.5$ and 12 K. The dotted line corresponds to the $V \propto |I|^3$ behavior. **c** The temperature dependence of α value obtained from the fitting (the black lines) in (**b**). Reprinted from [24], Copyright Author(s) 2019 Limited. Licensed under CC BY 4.0

which gives $a = 78.1\,\Omega$, $c = 1.93\,\Omega/T$, and $T_{c0} = 10.7\,K$. On the other hand, the finite resistance at the low-temperature region is explained in terms of the vortex flow above the Berezinskii-Kosterlitz-Thouless (BKT) transition temperature at which the binding of the vortex-antivortex pair realizes zero-resistance value. We fit the low-temperature region in terms of the BKT transition using the Halperin-Nelson formula [26–28].

$$R = R_0 \exp(-2b((T_{c0} - T)/(T - T_{BKT}))^{0.5}, \tag{5.3}$$

where R_0 and b are material parameters and T_{BKT} is the BKT transition temperature. The fitting gives $R_0 = 236\,\Omega$, $b = 1.5$, and $T_{BKT} = 8.0$ K. The well fitted curve by the Halperin-Nelson fomula confirms the two-dimensional nature of superconductivity as discussed in the literature [21].

Figure 5.2b shows the current-voltage (I-V) characteristics. The I-V characteristics are linear in the normal region ($T = 12$ K). The power α of I-V characteristics

$V = |I|^\alpha$ changes as a function of the temperature. In the log-log scale, the power α can be extracted from the slope of the curve. The extracted temperature dependence of the power is plotted in Fig. 5.2c. The jump in the power α from 1 to 3 is observed, which is characteristic of the two-dimensional superconductivity [28]. $\alpha = 3$ corresponds to the BKT transition temperature $T_{BKT} = 7.9$ K, which well coincides with the BKT transition temperature estimated from the temperature dependence of resistance value. The fitting of the temperature dependence of resistivity with the Berezinskii-Kosterlitz Thouless (BKT) transition [26–28] and the jump in the power law of current-voltage characteristics verify the two-dimensional nature of superconductivity [21], where the binding of the vortex-antivortex pairs realizes the zero-resistance state.

Next, we discuss the measurement procedure for the nonreciprocal transport. Since inversion symmetry is broken at the interface, the electrical voltage V under the in-plane B field is expressed as [13]:

$$V = R_0 I (1 + \gamma(\boldsymbol{B} \times z \cdot \boldsymbol{I})) \tag{5.4}$$

$$= R_0 I + \gamma R_0 B I^2. \tag{5.5}$$

The first term corresponds to the Ohm's law. When the second term is finite, the resistance is dependent on the direction of the current, namely the nonreciprocal transport appears. Because the second term is proportional to I^2, 2nd harmonic resistance $R^{2\omega} = \frac{R_0 \gamma B I}{\sqrt{2}}$ under ac current can detect the nonreciprocal transport. We fabricated a Hall bar of 100 μm width and measured the 2nd harmonic resistance using lock-in amplifiers [15, 17–20] under the ac current of $I = 200$ μA (this corresponds to $i = 2$ A/m in current density). Figure 5.1d shows the 2nd harmonic resistance as a function of the magnetic field. In the normal state ($T = 12$ K), the resistance value is negligibly small. In contrast, the B-linear 2nd harmonic resistance appears at $T = 9.5$ K below the superconducting onset temperature.

To reveal the origin of the sudden appearance of nonreciprocal transport below the onset of superconductivity, we study the characteristics of the 2nd harmonic signal in detail. In contrast to the first harmonic resistance, which is independent of the magnitude of the current, the 2nd harmonic resistance increases with current, as shown in Fig. 5.3a, . The temperature dependence of R^ω and γ is summarized in Fig. 5.3d, e. Here, γ value is derived from the slope of the magnetic field dependence of $R^{2\omega}/R^\omega$ as shown in Fig. 5.3c. To see the correspondence between the temperature dependence of the γ value, we categorize them into three regions; normal, intermediate, and superconducting regions. The 2nd harmonic voltage is not observed within the noise level in the normal region. The 2nd harmonic resistance vanishes with the disappearance of the first harmonic resistance in the superconducting region. In contrast, γ becomes finite in the intermediate region. Especially, γ value shows a divergent behavior towards the Berezinskii-Kosterlitz-Thouless (BKT) transition temperature T_{BKT}. The maximum of γ is 6.5×10^{-3} T^{-1}A^{-1} m at $T = 6.9$ K, which is more than 5 orders of magnitude larger than the reported value of $\gamma = 6 \times 10^{-8}$

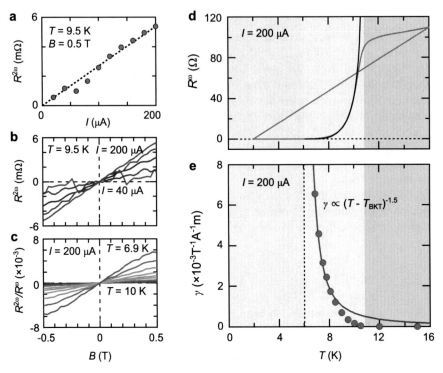

Fig. 5.3 Temperature and current-magnitude dependence of nonreciprocal transport. **a** The current magnitude dependence of 2nd harmonic resistance $R^{2\omega}$ at $T = 9.5$ K and $B = 0.5$ K. **b** The magnetic field dependence of 2nd harmonic resistance $R^{2\omega}$ at $T = 9.5$ K measured under $I = 40\text{–}200\,\mu\text{A}$. **c** The magnetic field dependence of $R^{2\omega}/R^{\omega}$ measured under $I = 200\,\mu\text{A}$ at $T = 6.9, 7.2, 7.5, 7.8, 8.1, 8.5, 9, 9.5$ and 10 K. **d** The temperature dependence of resistance measured under $I = 200\,\mu\text{A}$. The black curve is the fitting by the Berezinskii-Kosterlitz-Thouless (BKT) transition using Halperin-Nelson formula, $R = R_0 \exp(-2b((T_{c0} - T)/(T - T_{\text{BKT}}))^{0.5}$. $T_{c0} = 10.7$ K and $T_{\text{BKT}} = 6.0$ K are estimated from the fitting. The red, green and blue regions correspond to superconducting, intermediate and normal regions, respectively. **e** The temperature dependence of γ value under $I = 200\,\mu\text{A}$ as derived from c. The purple curve is the fitting with the formula $(T - T_{\text{BKT}}^0)^{-1.5}$. Reprinted from [24], Copyright Author(s) 2019 Limited. Licensed under CC BY 4.0

$T^{-1}A^{-1}m$ in TI without superconductivity [18]. The large enhancement of γ associated with superconductivity can be qualitatively explained as follows: The energy scale governing the system changes from the Fermi energy E_F of around a few hundred meV to the much smaller superconducting gap Δ of around 1 meV accompanied with the superconducting transition. Consequently, the relative energy scale of spin-orbit interaction and magnetic field gets larger, resulting in the enhancement of nonreciprocal transport [20].

A recent theoretical investigation [29] discusses a microscopic picture of nonreciprocal transport in a superconducting-proximitized TI surface state. According to this, the existence of spin-momentum locking renormalizes the supercurrent density with the application of electric current and in-plane magnetic field perpendicular to the current. Consequently, BKT transition temperature is modulated as $T_{BKT} = T_{BKT}^0(1 + \alpha BI)$. Substituting it to the standard BKT model, which is valid at around T_{BKT}, we get

$$R = R_1 \exp(-b(T - T_{BKT})^{-0.5}) \tag{5.6}$$

$$= R_1 \exp(-b(T - T_{BKT}^0)^{-0.5})(1 - 0.5\alpha b T_{BKT}^0(T - T_{BKT}^0)^{-1.5} BI) \tag{5.7}$$

Comparing the present equation with Eq. (5.5), the γ value is expected to diverge as $(T - T_{BKT}^0)^{-1.5}$ [29]. In fact, the fitting of γ by the formula $(T - T_{BKT}^0)^{-1.5}$ in Fig. 5.3e reproduces the divergence of nonreciprocal transport well towards T_{BKT}. Thus, we conclude that the nonreciprocal signal at a low B field region stems from the modulation of supercurrent density and that the superconducting nature of Bi_2Te_3/FeTe interface reflects the spin-momentum locking at the surface state.

Up to now, we have considered the 2nd harmonic signal at a low B field region. We next discuss the 2nd harmonic voltage at higher fields. Figure 5.4a, b show the B field dependence of resistance and 2nd harmonic resistance for $T = 9.5$ K and 10 K. At low magnetic fields, the 2nd harmonic signal is linear against the magnetic field. However, the signal undergoes a sign reversal at high magnetic fields. With the further application of the field, the 2nd harmonic voltage goes to zero associated with the breaking of superconductivity, as can been seen from the data at $T = 10$ K and $B = 14$ T. The B-T diagram for the first harmonic and 2nd harmonic signals is displayed in 2D plots of Fig. 5.4c, d, respectively. We can see that the 2nd harmonic resistance appears positively at low fields and negatively at high fields in the intermediate temperature region. To discuss the origin of the negative component, we investigate the magnetic field direction dependence of the 2nd harmonic voltage. Figure 5.4e, f show the in-plane and out-of-plane magnetic-field directional dependence of 2nd harmonic voltage, respectively. The signal becomes almost zero at $B||x$ and $B||z$. This confirms that the superconductivity reflects the polar symmetry at the interface [12, 15–18]. In xy plane, the signal shows $\sin\varphi$ dependence both at 2 T and 9 T, meaning that the signal scales with the y component of the magnetic field, as expected from the Eq. (5.5). Similarly, in yz plane, the magnetic field dependence shows $\sin\theta$ behavior at 2 T. At 9 T, the signal almost follows $\sin\theta$ with the same sign as at 2 T. Remarkably, abrupt sign reversal is observed at $\theta = 90°$ and $270°$, namely only when B is aligned almost in-plane. Interestingly, the negative component at high fields almost vanishes at $\theta = 70°$ as can be seen from Fig. 5.4g. This result implies that different origins govern the positive and the negative components. In the narrow θ range close to $\theta = 90°$ (or $270°$) at high fields, the in-plane magnetic flux lines along the y-direction are likely to coexist with the z-direction free vortices near above the BKT transition. Therefore, the interaction among these different vortices may bring about the new nonreciprocal charge transport. Further theoretical and experimental investigations are required to fully reveal the origins of the negative component at high fields. Figure 5.5 displays the angular dependence of 2nd harmonic resistance,

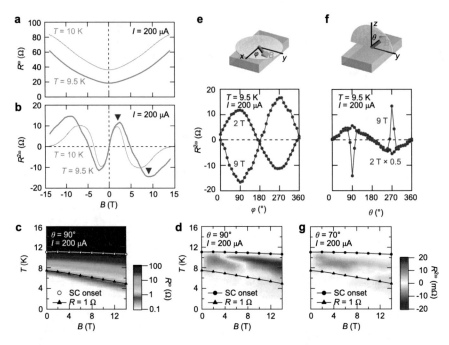

Fig. 5.4 Nonreciprocal transport under high magnetic fields. **a** The magnetic field dependence of resistance at $T = 9.5$ K (light blue) and $T = 10$ K (gray) measured under $I = 200\,\mu$A. **b** The magnetic field dependence of 2nd harmonic resistance at $T = 9.5$ K and $T = 10$ K measured under $I = 200\,\mu$A. The red and blue triangles represent the positive and negative peaks at around $B = 2$ T and $B = 9$ T, respectively. **c** The 2D plot of resistance at $\theta = 90°$ measured under $I = 200\,\mu$A in the plane of magnetic field and temperature. The superconducting onset and $R = 1\,\Omega$ are shown in circle and triangle, respectively. **d** The 2D plot of 2nd harmonic resistance at $\theta = 90°$ measured under $I = 200\,\mu$A in the plane of magnetic field and temperature. **e** The in-plane magnetic-field direction dependence of 2nd harmonic resistance $B = 2$ T (red) and $B = 9$ T (blue) within xy plane measured under $I = 200\,\mu$A. **f** The out-of-plane magnetic-field direction dependence of 2nd harmonic resistance $B = 2$ T (red) and $B = 9$ T (blue) within zy plane measured under $I = 200\,\mu$A. **g** The 2D plot of 2nd harmonic resistance at $\theta = 70°$ measured under $I = 200\,\mu$A in the plane of magnetic field and temperature. Reprinted from [24], Copyright Author(s) 2019 Limited. licensed under CC BY 4.0

which helps to understand the discussion in the main text. In xy-plane (Fig. 5.5a–c), the signal shows $\sin\varphi$ dependence for all the magnetic field region. On the other hand, in yz-plane (Fig. 5.5d–f), the signal does not follow $\sin\theta$ dependence. It can be seen that the sign reversal appears only in the small pocket, where the high magnetic field is aligned almost perfectly to the in-plane direction.

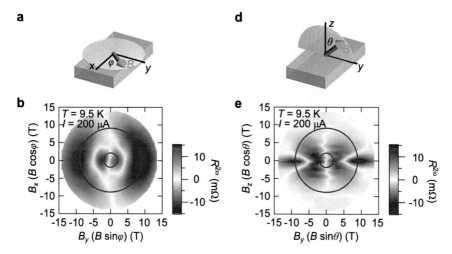

Fig. 5.5 **a** Schematic drawing of the magnetic field direction dependent measurement in xy-plane. **b** The color plot of 2nd harmonic resistance under magnetic field in xy-plane. The abscissa (vertical) axis corresponds to y (x) component of the magnetic field B_y (B_x). The measurement is done at $T = 9.5$ K and $I = 200$ μA . **c** The magnetic field direction dependence of 2nd harmonic resistance at $B = 2$ T (red) and $B = 9$ T (blue) as indicated in the circles in (**b**). **d–f**, The same as a-c for yz-plane. Reprinted from [24], Copyright Author(s) 2019 Limited. Licensed under CC BY 4.0 (Color figure in online)

5.3 Summary

The observation of sizeable nonreciprocal signal in the two-dimensional superconductivity of Bi_2Te_3/FeTe interface demonstrates that the Cooper pairs reflect the broken inversion symmetry originating from the surface spin-momentum locking of a TI. The nonreciprocal transport at the low magnetic field is attributed to the modulation of supercurrent density under the in-plane magnetic field and current. The negative component at high field appears only when the magnetic field is aligned to in-plane direction. Further study is required to reveal the origin of the nonreciprocal effect completely. Because of the coupling between superconductivity and topological surface state, the interface superconductivity between Bi_2Te_3 and FeTe serves as a desirable platform to study TSC and the associated formation of Majorana fermion by tuning the Fermi level to the surface state [30]. The nonreciprocal transport measurement can also be applied to other TSC candidates such as $LaAlO_3$/$SrTiO_3$ interface [31], superconducting-proximity coupled Rashba wire [5, 6] and topological surfaces [8, 10, 11, 32] , which will help to discuss the pairing symmetry of the superconductors.

References

1. Kitaev A (2006) Anyons in an exactly solved model and beyond. Ann Phy 321(1):2–111
2. Wilczek F (2009) Majorana returns. Nat Phys 5(9):614–618
3. Alicea J (2012) New directions in the pursuit of majorana fermions in solid state systems. Rep Prog Phys 75(7):076501
4. Oreg Y, Refael G, Von Oppen F (2010) Helical liquids and majorana bound states in quantum wires. Phys Rev Lett 105(17):177002
5. Mourik V, Zuo K, Frolov SM, Plissard SR, Bakkers EPAM, Kouwenhoven LP (2012) Signatures of majorana fermions in hybrid superconductor-semiconductor nanowire devices. Science 336(6084):1003–1007
6. Nadj-Perge S, Drozdov IK, Li J, Chen H, Jeon S, Seo J, MacDonald AH, Bernevig BA, Yazdani A (2014) Observation of majorana fermions in ferromagnetic atomic chains on a superconductor. Science 346(6209):602–607
7. Fu L, Kane CL (2008) Superconducting proximity effect and majorana fermions at the surface of a topological insulator. Phys Rev Lett 100(9):096407
8. Wiedenmann J, Bocquillon E, Deacon RS, Hartinger S, Herrmann O, Klapwijk TM, Maier L, Ames C, Brüne C, Gould C et al 4 π-periodic Josephson supercurrent in HgTe-based topological Josephson junctions. Nat Commun 7(1):1–7, 2016
9. He QL, Pan L, Stern AL, Burks EC, Che X, Yin G, Wang J, Lian B, Zhou Q, Choi ES et al (2017) Chiral majorana fermion modes in a quantum anomalous hall insulator–superconductor structure. Science 357(6348):294–299
10. Wang M-X, Liu C, Xu J-P, Yang F, Miao L, Yao M-Y, Gao CL, Shen C, Ma X, Chen X et al (2012) The coexistence of superconductivity and topological order in the bi2se3 thin films. Science 336(6077):52–55
11. Su-Yang X, Alidoust N, Belopolski I, Richardella A, Liu C, Neupane M, Bian G, Huang S-H, Sankar R, Fang C et al (2014) Momentum-space imaging of cooper pairing in a half-dirac-gas topological superconductor. Nat Phys 10(12):943–950
12. Ideue T, Hamamoto K, Koshikawa S, Ezawa M, Shimizu S, Kaneko Y, Tokura Y, Nagaosa N, Iwasa Y (2017) Bulk rectification effect in a polar semiconductor. Nat Phys 13(6):578–583
13. Rikken GLJA, Wyder P (2005) Magnetoelectric anisotropy in diffusive transport. Phys Rev Lett 94(1):016601
14. Olejník K, Novák V, Wunderlich J, Jungwirth T (2015) Electrical detection of magnetization reversal without auxiliary magnets. Phys Rev B 91(18):180402
15. Avci CO, Garello K, Ghosh A, Gabureac M, Alvarado SF, Gambardella P (2015) Unidirectional spin hall magnetoresistance in ferromagnet/normal metal bilayers. Nat Phys 11(7):570–575
16. Yasuda K, Tsukazaki A, Yoshimi R, Takahashi KS, Kawasaki M, Tokura Y (2016) Large unidirectional magnetoresistance in a magnetic topological insulator. Phys Rev Lett 117(12):127202
17. Lv Y, Kally J, Zhang D, Lee JS, Jamali M, Samarth N, Wang J-P (2018) Unidirectional spin-hall and rashba-edelstein magnetoresistance in topological insulator-ferromagnet layer heterostructures. Nat Commun 9(1):1–7
18. He P, Zhang SS-L, Zhu D, Liu Y, Wang Y, Yu J, Vignale G, Yang H (2018) Bilinear magnetoelectric resistance as a probe of three-dimensional spin texture in topological surface states. Nat Phys 14(5):495–499
19. Qin F, Shi W, Ideue T, Yoshida M, Zak A, Tenne R, Kikitsu T, Inoue D, Hashizume D, Iwasa Y (2017) Superconductivity in a chiral nanotube. Nat Commun 8(1):1–6
20. Wakatsuki R, Saito Y, Hoshino S, Itahashi YM, Ideue T, Ezawa M, Iwasa Y, Nagaosa N (2017) Nonreciprocal charge transport in noncentrosymmetric superconductors. Sci Adv 3(4):e1602390
21. He QL, Liu H, He M, Lai YH, He H, Wang G, Law KT, Lortz R, Wang J, Sou IK (2014) Two-dimensional superconductivity at the interface of a bi 2 te 3/fete heterostructure. Nat Commun 5(1):1–8
22. Zhang H, Liu C-X, Qi X-L, Dai X, Fang Z, Zhang S-C (2009) Topological insulators in bi 2 se 3, bi 2 te 3 and sb 2 te 3 with a single dirac cone on the surface. Nat Phys 5(6):438–442

23. Bao W, Qiu Y, Huang Q, Green MA, Zajdel P, Fitzsimmons MR, Zhernenkov M, Chang S, Fang M, Qian B et al (2009) Tunable ($\delta \pi$, $\delta \pi$)-type antiferromagnetic order in α-fe (te, se) superconductors. Phys Rev Lett 102(24):247001

24. Yasuda K, Yasuda H, Liang T, Yoshimi R, Tsukazaki A, Takahashi KS, Nagaosa N, Kawasaki M, Tokura Y (2019) Nonreciprocal charge transport at topological insulator/superconductor interface. Nat Commun 10(1):1–6

25. Aslamasov LG, Larkin AI (1968) The influence of fluctuation pairing of electrons on the conductivity of normal metal. Phys Lett A 26(6):238–239

26. Berezinsky VL (1972) Destruction of long-range order in one-dimensional and two-dimensional systems possessing a continuous symmetry group. ii. quantum systems. Zh Eksp Teor Fiz 61:610

27. Kosterlitz JM, Thouless DJ (1973) Ordering, metastability and phase transitions in two-dimensional systems. J Phys C: Solid State Phys 6(7):1181

28. Halperin BI, Nelson DR (1979) Resistive transition in superconducting films. J Low Temp Phys 36(5-6):599–616

29. Hoshino S, Wakatsuki R, Hamamoto K, Nagaosa N (2018) Nonreciprocal charge transport in two-dimensional noncentrosymmetric superconductors. Phys Rev B 98(5):054510

30. Zhang J, Chang C-Z, Zhang Z, Wen J, Feng X, Li K, Liu M, He K, Wang L, Chen X et al (2011) Band structure engineering in (bi 1–x sb x) 2 te 3 ternary topological insulators. Nat Commun 2(1):1–6

31. Mohanta N, Taraphder A (2014) Topological superconductivity and majorana bound states at the laalo3/srtio3 interface. EPL (Europhysics Letters) 108(6):60001

32. Xu J-P, Wang M-X, Liu ZL, Ge J-F, Yang X, Liu C, Xu ZA, Guan D, Gao CL, Qian D et al (2015) Experimental detection of a majorana mode in the core of a magnetic vortex inside a topological insulator-superconductor bi 2 te 3/nbse 2 heterostructure. Phys Rev Lett 114(1):017001

Chapter 6
Summary

In the final chapter, we summarize the result obtained in each chapter and discuss their imprecation for the future study of a topological insulator with broken symmetry.

In Chap. 3, we discussed the high-temperature realization of QHE and QAHE and the associated control/manipulation of CES at the domain wall. First, by making heterostructures of BST and CBST using molecular beam epitaxy (MBE) technique, the realization temperatures of QHE and QAHE is raised from up to about $T = 1$ K, which is much higher than the case of single-layer BST and CBST. The higher temperature realization of quantum (anomalous) Hall effect is attributed to the realization of surface state with smaller magnetic disorder. The formation of complex heterostructure, including magnetic proximity effect is expected to further increase the realization temperature of QAH and QAHE [1–3]. That will expand the available experimental techniques for obtaining further physical insights and opening a pathway towards the application of a dissipationless electronic channel. The realization of high-temperature QHE and QAHE TI-based heterostructures and superlattices will also provide a new platform in exploring new functionalities and exotic topological phases such as Weyl semimetal [4, 5].

Utilizing the higher temperature QAHE, we next confirmed the existence of CESs at the magnetic DW. We designed and fabricated the magnetic domains on QAH state with the tip of a magnetic force microscope, and proved the existence of the chiral one-dimensional edge conduction along the prescribed DWs through transport measurements. The proof-of-concept devices based on reconfigurable CESs and Landauer-Büttiker formalism are also realized for multiple-domain configurations with well-defined DW channels. The experimental methods to control the magnetization configuration in a local, rapid, and reconfigurable manner will introduce the CESs into applicable devices in the future. One way to achieve this goal is to use the spin-orbit torque magnetization switching by the spin accumulation at the surface state [6, 7]. Another way is to use a magneto-optical recording, namely optical writing of magnetization through the modulation of the coercivity by heating

© Springer Nature Singapore Pte Ltd. 2020   93
K. Yasuda, *Emergent Transport Properties of Magnetic Topological Insulator Heterostructures*, Springer Theses,
https://doi.org/10.1007/978-981-15-7183-1_6

[8]. The effective control of the magnetic configuration will lead to the low power-consumption CES-based logic and memory devices [9–12], and quantum information processing [13–15] in the future.

In Chap. 4, we investigated the spintronics functionalities of magnetic TI, CBST. Because of the strong coupling between the spin-momentum locked surface state and local magnetization, rich spintronic phenomena appear in a magnetic TI. Especially, magnetic TI is shown to yield a larger spintronic response compared with NM/FM heterostructures. First, we studied unidirectional magnetoresistance (UMR) or current-direction dependence magnetoresistance in magnetic TI. UMR is shown to be several orders of magnitude larger than in other reported systems [16–18]. From the detailed temperature and magnetic field dependence, the origin of UMR is identified to be asymmetric magnon scattering. Similar nonlinear transport is observed in a transverse resistance when the magnetic field is parallel to the current, namely the current-nonlinear Hall effect. It was revealed that the current-nonlinear Hall effect also originates from asymmetric magnon scattering. The asymmetric magnon scattering is revealed to lead to the overestimation of spin Hall angle by current-nonlinear Hall effect. Next, current-induced magnetization switching is realized in a magnetic TI. The required current for switching is shown to be about one order of magnitude smaller than other systems. Finally, we demonstrated the skyrmion formation from the surface state of magnetic TI. The spin-momentum locking results in the appearance of finite DM interaction at the surface state, which leads to the formation of Néel type skyrmion. The current injection to skyrmion also resulted in the motion of skyrmion opposite to the current direction. The room temperature realization of these functionalities in a magnetic TI will open up a route towards the practical application of these effects.

In Chap. 5, we measured the transport property of superconducting proximity coupled topological surface state at the interface of Bi_2Te_3/FeTe. The nonreciprocal transport is shown to be largely enhanced with superconductivity, meaning that the Cooper pairs reflect the broken inversion symmetry of the surface spin-momentum locking. Thus, by tuning the Fermi level to the surface state [19], the interface superconductivity between Bi_2Te_3 and FeTe is expected to serve as a desirable platform to study TSC and the associated formation of Majorana fermion. The nonreciprocal transport measurement can be widely applied to other TSC candidates, which will help to discuss the pairing symmetry of the superconductors. The confirmation of Majorana fermion at the vortex core and the sample edge in Bi_2Te_3/FeTe remains as a future issue.

To summarize, we uncovered emergent transport properties and versatile functionalities of magnetic and superconducting proximity coupled topological insulator heterostructures. The development of heterostructure engineering, device fabrication, and measurement technique of topological insulators with broken symmetry will lead to the further discovery of novel functionalities in the future.

References

1. Tang C, Chang C-Z, Zhao G, Liu Y, Jiang Z, Liu C-X, McCartney MR, Smith DJ, Chen T, Moodera JS et al (2017) Above 400-k robust perpendicular ferromagnetic phase in a topological insulator. Sci Adv 3(6):e1700307, 2017
2. Mogi M, Nakajima T, Ukleev V, Tsukazaki A, Yoshimi R, Kawamura M, Takahashi KS, Hanashima T, Kakurai K, Arima T-H et al (2019) Large anomalous hall effect in topological insulators with proximitized ferromagnetic insulators. Phys Rev Lett 123(1):016804
3. Watanabe R, Yoshimi R, Kawamura M, Mogi M, Tsukazaki A, Yu XZ, Nakajima K, Takahashi KS, Kawasaki M, Tokura Y (2019) Quantum anomalous hall effect driven by magnetic proximity coupling in all-telluride based heterostructure. Appl Phys Lett 115(10):102403
4. Burkov AA, Balents L (2011) Weyl semimetal in a topological insulator multilayer. Phys Rev Lett 107(12):127205
5. Qi X-L, Hughes TL, Zhang S-C (2008) Topological field theory of time-reversal invariant insulators. Phys Rev B 78(19):195424
6. Mellnik AR, Lee JS, Richardella A, Grab JL, Mintun PJ, Fischer MH, Vaezi A, Manchon A, Kim E-A, Samarth N et al (2014) Spin-transfer torque generated by a topological insulator. Nature 511(7510):449–451
7. Yasuda K, Tsukazaki A, Yoshimi R, Kondou K, Takahashi KS, Otani Y, Kawasaki M, Tokura Y (2017) Current-nonlinear hall effect and spin-orbit torque magnetization switching in a magnetic topological insulator. Physical review letters 119(13):137204
8. Yeats AL, Mintun PJ, Pan Y, Richardella A, Buckley BB, Samarth N, Awschalom DD (2017) Local optical control of ferromagnetism and chemical potential in a topological insulator. Proc Nat Acad Sci 114(39):10379–10383
9. Catalan G, Seidel J, Ramesh R, Scott JF (2012) Domain wall nanoelectronics. Rev Mod Phys 84(1):119
10. Parkin SSP, Hayashi M, Thomas L (2008) Magnetic domain-wall racetrack memory. Science 320(5873):190–194
11. Pramey U, Yaroslav T (2016) Domain wall in a quantum anomalous hall insulator as a magnetoelectric piston. Phys Rev B 94(2):020411
12. Yaroslav T, Daniel L (2012) Thin-film magnetization dynamics on the surface of a topological insulator. Phys Rev Lett 108(18):187201
13. Xiao-Liang Q, Shou-Cheng Z (2011) Topological insulators and superconductors. Rev Mod Phys 83(4):1057
14. Fu L, Kane CL (2009) Probing neutral majorana fermion edge modes with charge transport. Phys Rev Lett 102(21):216403
15. Akhmerov AR, Nilsson J, Beenakker CWJ (2009) Electrically detected interferometry of majorana fermions in a topological insulator. Phys Rev Lett 102(21):216404
16. Olejník K, Novák V, Wunderlich J, Jungwirth T (2015) Electrical detection of magnetization reversal without auxiliary magnets. Phys Rev B 91(18):180402
17. Avci CO, Garello K, Ghosh A, Gabureac M, Alvarado SF, Gambardella P (2015) Unidirectional spin hall magnetoresistance in ferromagnet/normal metal bilayers. Nat Phys 11(7):570–575
18. Avci CO, Garello K, Mendil J, Ghosh A, Blasakis N, Gabureac M, Trassin M, Fiebig M, Gambardella P (2015) Magnetoresistance of heavy and light metal/ferromagnet bilayers. Appl Phys Lett 107(19):192405
19. Jinsong Z, Cui-Zu C, Zuocheng Z, Jing W, Xiao F, Kang L, Minhao L, Ke H, Lili W, Xi C et al (2011) Band structure engineering in (bi 1–x sb x) 2 te 3 ternary topological insulators. Nat Commun 2(1):1–6

Curriculum Vitae

Kenji Yasuda
Postdoctoral Fellow
Department of Physics, Massachusetts Institute of
Technology
Cambridge, Massachusetts 02139, United States of
America
e-mail: yasuda@mit.edu
Web: http://sites.google.com/view/kenjiyasuda/

Appointments

- JSPS Overseas Research Fellowships, Massachusetts Institute of Technology (Apr 2019 –)
- Postdoctoral Associate with Prof. Pablo Jarillo-Herrero, Massachusetts Institute of Technology (Oct 2018–Mar 2019)

Education

- Doctor of Philosophy (Ph.D.), Dept. of Applied Physics, Graduate School of Engineering, University of Tokyo, advisor Prof. Yoshinori Tokura (Apr 2014– Sep 2018)
- Master of Science (MSc) Dept. of Applied Physics, Graduate School of Engineering,
  University of Tokyo, advisor Prof. Yoshinori Tokura (Apr 2014–Mar 2016)

© Springer Nature Singapore Pte Ltd. 2020
K. Yasuda, *Emergent Transport Properties of Magnetic Topological*
Insulator Heterostructures, Springer Theses,
https://doi.org/10.1007/978-981-15-7183-1

- Bachelor of Science (BS) Dept. of Physics, Faculty of Science, University of Tokyo (Apr 2010–Mar 2014)

Service

- Reviewer for Nature, Nature Nanotechnology, Nature Electronics, Nature Communications, Nano Letters, npj Quantum Materials

Teaching

- Advisor for a graduate student at Massachusetts Institute of Technology for a graduate research project (Oct 2018 –)
- Advisor for an undergraduate visiting student at Massachusetts Institute of Technology for an undergraduate research project (Sep 2019–May 2020)
- Advisor for an undergraduate student at the University of Tokyo for an undergraduate research project (Apr 2017–Feb 2018)

Academic Awards and Fellowships

- President's award of University of Tokyo (Mar 2019)
- Dean's Award in the Graduate School of Engineering, University of Tokyo (Mar 2019)
- 9th JSPS Ikushi Prize (Mar 2019)
- CEMS Symposium on Trends in Condensed Matter Physics, Science Poster Prize (Nov 2017)
- International workshop on nano-spin conversion science & quantum spin dynamics (NSCS-QSD), Poster award (Oct 2016)
- Japan Society for the Promotion of Science Research Fellow (DC1) (Apr 2016–Mar 2020)
- Dean's Award in the Graduate School of Engineering, University of Tokyo (Mar 2016)
- Shoji Tanaka prize (master's thesis award),Graduate School of Engineering, University of Tokyo (Mar 2016)

Research Statement

My research aim is to study emergent physical properties of the heterostructures of low-dimensional systems. It has long been discussed in condensed matter physics that the control of the atomic arrangement and interface in a nanoscale material will realize novel quantum phases different from bulk crystals. The recent advances in the thin film growth, van der Waals heterostructure, and nanofabrication technology brought such imagination to reality. My primary interest is to find emergent electronic states such as topological phenomena, magnetism, ferroelectricity, and superconductivity and to study the interplay between them employing atomic engineering. In addition to basic characterization such as transport measurement, the simultaneous study of nanoscale domain configuration allows us to connect the microscopic structure with macroscopic physical property. Furthermore, the structural degrees of freedom and low-dimensionality of thin films and flakes enable the control of their quantum phases through minute external stimulus. My goal is to discover, understand, and control the emergent physical property of low-dimensional heterostructures for constructing novel building blocks of next-generation electronic devices.

Printed in the United States
by Baker & Taylor Publisher Services